The ISO 9000 Audit

A Guide to Understanding and Preparing for Quality System Audits

Daren P. Unger

Raymond J. Murphy

Part 1
The Audit for Managers
Provides assistance for managers in preparing for both internal and external audits

Part 2
Internal Auditor Training
Provides an understanding of the concepts and techniques needed to implement an effective internal audit program and provides the material to effectively train internal auditors

Part 3
Reference Material
Provides reference material that is useful in establishing an effective internal audit program

 Government Institutes
a Division of ABS Group Inc.
Rockville, Maryland

Government Institutes, a Division of ABS Group Inc.
4 Research Place, Rockville, Maryland 20850, USA
Phone: (301) 921-2300
Fax: (301) 921-0373
Email: giinfo@govinst.com
Internet: http://www.govinst.com

Library of Congress Cataloging-in-Publication Data

Murphy, Raymond J., 1938–
     The ISO 9000 audit: a guide to understanding and preparing for quality system audits / Raymond J. Murphy,
Darren P. Unger.
          p.   cm.
     Includes bibliographical references.
     ISBN: 0-86587-657-6 (pbk.)
     1. ISO 9000 Series Standards—Auditing.   I. Unger, Darren P.   II. Title.
     TS156.6.M88   1999
     658.5'2—dc21   99-20340
                                                            CIP

Printed in the United States of America

SUMMARY TABLE OF CONTENTS

TABLE OF CONTENTS

PART I - THE AUDIT FOR MANAGERS

This chapter will explain that audits are management tools for verifying conformance with stated requirements. The chapter will also highlight popular misconceptions about audits.

To prepare for an audit, the auditee needs to know why audits take place and what the differences are between internal and external audits. This chapter examines the driving factors that lead an organization to conduct internal audits and the driving factors behind external audits. The chapter prepares the manager to deal with the most commonly asked question: "Why do we have to be audited?"

Chapter 3 **Types of Audits** **13**

The type of audit performed will depend on the objective of the audit. Chapter 3 teaches the manager about the various types of audits the organization can expect to face in the future. Knowing the differences will help managers and staffs prepare effectively.

Chapter 4 **Understanding and Meeting the Requirements** **17**

This chapter discusses items that must be considered by the manager before the audit. The chapter centers on identifying the quality system requirements for the department and conducting a self-assessment to determine the department's preparedness for the actual audit. This chapter also touches on the training of personnel to carry out the processes conducted by the department.

Chapter 5   Setting the Stage For An Audit **30**

This chapter discusses audit preparation items that are outside the realm of compliance but which are important to the success of the audit. Topics covered include organizational charts, housekeeping, and the availability of staff.

Chapter 6   Conducting the Audit **35**

To ensure success in an audit, managers and staff alike need to understand the mechanics of the audit process. This chapter takes the manager through the audit investigation process from the opening meeting through the questioning phase to the closing meeting. Additionally, sample procedures and process instructions for conducting internal audits are given.

Chapter 7   Coping with Audit Fear **42**

Fear is a problem with which all of those who are audited must learn to deal, especially those being audited for the first time. This chapter provides an understanding of the causes of fear and how to deal with it before, during, and after the audit.

This chapter presents the techniques an auditor will use to conduct an audit. An understanding of these techniques will allow managers to anticipate how an audit will progress, and allow proper preparation of the staff.

This chapter presents effective communications techniques. How the person(s) being audited answers audit questions can play a significant role in the outcome of the audit. Managers and staff alike must master the techniques of effective communication in order to succeed in the audit process.

The findings of an audit are documented in an audit report that allows the auditee to follow-up and take corrective and preventive action. This module discusses the audit report and necessary steps that the auditee must take to resolve the nonconformances and observations.

PART 2 – INTERNAL AUDITOR TRAINING

Why audit? What is an internal audit? Are there other types of audits? This chapter explains why the internal audit is the most appropriate form for measuring a new quality system and for maintaining an existing quality system.

This chapter looks at audit planning and how to set the scope of an audit. It explains how to select an audit team, how to review documented requirements and reports before conducting an audit, how and when to notify an auditee of an upcoming audit, and discusses the mechanics of a pre-audit team meeting.

Chapter 13 **Development of Audit Checklists** **85**

An audit checklist is the key to a well organized audit. Chapter 3 guides auditors in the development of effective audit checklists. The purpose of the audit checklist is discussed, as well as how to identify processes and requirements that should be included, how to develop meaningful, objective questions, and how to make audit checklists user-friendly.

Chapter 14 **Audit Techniques** **90**

This chapter presents investigative skills and techniques every auditor must master to ensure a successful internal audit. Skills include effective communication, how to separate fact from opinion, and how to take good notes. Techniques include data sampling, how to use an audit team effectively, and how to deal with confrontation or evasion.

Chapter 17 Auditor Qualities and Certification

Patience and objectivity are two characteristics of an ideal internal auditor. This chapter provides tips on how to develop these and other essential characteristics.

PART 3 - REFERENCE MATERIAL

LIST OF FIGURES

PREFACE

With growing worldwide competition for resources and customers, more and more companies are being forced to review their management systems and processes with the consideration of making them as cost-effective and efficient as possible. This involves the elimination of non-value added steps from those processes, determination of process deficiencies and the filling in of the gaps where deficiencies are found. This system and process review must include process management, that is, the way work is done. Process management is one definition of quality.

Process management, whether developed within a company or imposed by an external organization, is the means through which a business takes control of its processes and a window through which others can see that uniform quality is the most likely outcome. The focus on processes and the understanding of audits are two of the keys to success, leading to certification of the process management systems. In all process management systems, which are ISO 9000 based, the audit is the primary technique for measuring a company's ability to do what it says it does.

Part 1 of this book is specifically designed to assist the manager in preparing for both internal and external audits. In addition, Part 1 will provide the manager with an understanding of the concepts and techniques used in the audit process.

Part 2 of this book addresses internal auditor training and is specifically designed to provide the reader with an understanding of the concepts and techniques needed to carry out an effective internal audit. Having the ability to conduct internal audits in a planned manner, using trained personnel, will assist the organization in preparing for successful certification of their management system. In addition, the contents of Part 2 can be used to effectively train internal auditors.

Part 3 of this book contains reference material that is useful in establishing an effective internal audit program.

ABOUT THE AUTHORS

Darren P. Unger is the Manager of Quality Documentation and a Certified Lead Auditor for the American Bureau of Shipping. He was instrumental in the establishment of the worldwide Internal Audit Program for the American Bureau of Shipping involving over 60 auditors, 120 offices and 90 countries. He has, as a key team member, managed the development and implementation of ISO 9000 compliant Quality Systems in Engineering, Inspection, Finance, Human Resources and the Information Technology fields. Mr. Unger has a broad range of auditing experience in different disciplines, having conducted over 200 internal audits in 40 countries worldwide. These audits ranged from single person offices to the audits of large engineering and corporate offices where Mr. Unger led the audit teams. Mr. Unger has taught the principles of auditing to both prospective auditors and management in preparation for their respective roles during internal and external certification audits. He has authored auditing training courses for use by the ABS Academy. Mr. Unger's extensive auditing experience uniquely qualifies him for the subject of this book.

Mr. Unger is a resident of Hoboken, New Jersey and holds both Bachelors and Masters of Engineering degrees from Stevens Institute of Technology.

Raymond J. Murphy is the Director of Quality for the American Bureau of Shipping. He is a seasoned continuous improvement process leader and practitioner with hands-on experience in planning/implementing management systems and processes in challenging environments; orchestrating/sustaining process improvement through the involvement of people; strategic planning; managing all phases of operations including Engineering, Research and Development, Manufacturing, Quality Assurance, Project Management, Material Control, Engineering Support, Document Control, and Technical Publications. He directed the establishment of Total Quality Improvement Processes in several international organizations/industries engaged in technical services/consulting; planning analysis; and process and utility. He served in several global, technical organizations as engineer, manager, senior executive, and consultant. He is a skilled facilitator and communicator with the ability to identify and resolve complex problems using a wide range of tools and techniques. He is known for his reputation as a process improvement pioneer and leader who works effectively with all levels of people and "walks the talk." Mr. Murphy has conducted over 300 internal audits on a worldwide basis, taught and qualified over 100 internal auditors and developed a variety of check sheets for internal audits. This combination of Mr. Murphy's background uniquely qualifies him in the subject of this book.

Mr. Murphy served as a Malcolm Baldrige National Quality Award Examiner for three years and has received the Instrument Society of America's Excellence in Documentation Award. He resides in Conroe, Texas, with his wife and three daughters.

PART 1

THE AUDIT FOR MANAGERS

*Provides assistance for
managers in preparing for both
Internal and External audits*

Chapter 1

WHAT AN AUDIT IS — AND IS NOT

When faced with an impending audit, most people can only speculate about what an audit really is. In this chapter we discuss what a quality audit really is and, on the other side, what it really is not. By knowing the difference, managers can prepare for the event and reduce the associated stress felt by the employees.

What Is an Audit?

> ◆ A management tool to verify conformance to quality system requirements
> ◆ A method by which to determine compliance with an external standard
> ◆ An objective means for verifying the organization's activities
> ◆ A method to identify opportunities to improve:
>   • quality system
>   • how the organization carries out its processes

One way to define the word "audit" would be to open a dictionary. Webster's Ninth New Collegiate Dictionary offers the following definition:

> Audit\'od-et\ n [ME, fr. L. auditus act of hearing, fr. Auditus, pp.]
> (15c) 1.a: a formal examination of an organization's or individual's
> accounts or financial situation b: the final report of an audit 2: a
> methodical examination and review – auditable \-e-bel\adj
>
> *Webster's Ninth Collegiate Dictionary*

Although this definition provides insight into what an audit is, the reader will still have questions about how this definition relates to a quality audit. The next place one might look would be to the ISO standards, particularly the Quality - Vocabulary standard ISO 8402-1986, which defines a quality audit as follows:

> Quality Audit – A systematic and independent examination to determine whether quality activities and related results comply with planned arrangements and whether these arrangements are implemented effectively and are suitable to achieve objectives.
>
> *Quality – Vocabulary ISO 8402-1986*

With this definition, the objectives of a quality audit become somewhat clearer, although most people will still struggle with details of interpretation and what is considered effective and suitable.

The manager and all employees need to view the quality audit as a process designed to improve the way work is done.

Every company carries out processes for every facet of work it does, and these processes are followed on a daily basis. Within organizations that have chosen to develop and implement a quality system, the requirements of each of the processes they carry out are documented. The quality audit measures conformance to these documented requirements on an objective and independent basis.

The quality audit from this perspective is a management tool that employs a systematic method of questioning and data collection to verify that the organization is actually doing what it claims it does. These audits may be conducted either by individuals within the organization or by external bodies.

From the manager's point of view, the audit is the most powerful tool available to find opportunities where improvements can be made in their processes. This message of improvement must be understood as the key concept behind the reason to audit and be audited. It is the manager's responsibility in the audit process to make the staff members aware of this message.

The basis of a quality audit, internal or external, is the organization's own quality system requirements, which may be in accordance with external standards such as the ISO 9000 series standard.

What Management Will Learn During an Audit

Management should be able to learn the following from an audit of their organization:

◆ Level of conformance to the requirements
◆ If the employees understand the quality system
◆ How well the system has been implemented
◆ If the quality system is being maintained at all levels
◆ If the quality system meets the needs of:
  ● the organization
  ● the customers
  ● the standards
◆ If the quality system provides effective controls to ensure conformance

Organizations that pursue certification to one of the ISO 9000 series quality standards design their own systems around the requirements of that standard. Therefore, an internal or external audit of the organization's quality system will verify that their system complies with the relevant clauses of the ISO 9000 standard. Similarly, when a company or organization does business that requires compliance to a different specific standard, compliance can be obtained by incorporating the relevant aspects of the standard into the internal quality system requirements.

What an Audit Is Not

Half the picture is knowing what an audit is. It is equally important for managers and employees to understand what an audit is not. Many people have the misconception that an audit is a necessary evil, an in-depth search to find nonconformances, a method of finding employees who are not doing their jobs, or a substitute for inspection. We will look at each of these misconceptions in detail in order to understand why these misconceptions must be eliminated from the work place.

An Audit Is Not a Search For Nonconformances

Contrary to popular belief, an auditor's mission is not to find as many things wrong in an organization's quality system as possible. The objective the auditor has when conducting the audit is to verify that what a company says it does it is actually doing, i.e., verifying compliance. The auditor, whether internal or external, just wants to see that the auditees' system has been developed, implemented and maintained in accordance with the stated requirements. The mindset of an auditor is to help the organization determine where it does what it says it does and where it does not. This said, managers and employees alike must understand that auditors will invariably find some nonconformances during an audit. Realizing this, the auditee must view the auditor as someone available to help identify opportunities for improving the company's quality system.

> ◆ An auditor's mission is to verify conformance through: questioning, objective and factual evidence
>
> ◆ Auditors ask themselves "Is the organization doing what it claims it is doing?" not "What is the organization doing wrong?"
>
> ◆ Auditees must realize that the auditor will find nonconformances, but not by looking for them

An Audit Is Not a Search for Wrongdoers

An audit is not a tool or weapon by which the organization identifies and punishes individuals or departments. Audits, whether internal or external, are used to determine how well the organization complies with stated requirements, and never used to satisfy personal vendettas. Management must understand that audit results are used to measure compliance and stimulate continuous improvement of the system. The credibility of the quality system and the auditing process depends on audits being perceived, by employees on all levels, as fair and impartial.

> ◆ An audit is not a weapon to identify wrongdoers
>
> ◆ An audit is not a method by which wrongdoers are punished
>
> ◆ An audit is used to address problems with the quality system, not with the employees
>
> ◆ Training and corrective/preventive action, rather than punishment, makes the quality system work

An Audit Is Not a Necessary Evil

Many times, management and staff alike view the auditing process as a necessary evil without which the organization cannot do business. Each organization chooses how and with whom it should do business and, therefore, can choose whether or not to be audited. As part of doing business, many organizations choose to have their methods of conducting business viewed by another organization to demonstrate publicly that what they say they are doing they actually are doing. Most organizations that choose to be audited view the audit process not as a necessary evil, but as an opportunity to strengthen weak areas. This opportunity to improve, if seized by the management and employees, pays benefits in higher profits, higher productivity, and makes the organization more efficient. Everyone from the top of the organization to the bottom will profit from the improvements made.

> ◆ Organizations choose to be audited for compliance with requirements
> ◆ The choice to be audited may be based on:
>   • whom the organization wants to do business with
>   • the organization's desire to determine where its quality system stands
>   • the organization's aspiration to external certification
> ◆ The decision to be audited is solely the choice of the organization

An Audit Is Not a Substitute for Inspection

Some managers view audits as a way to eliminate inspections of goods or services produced by the organization. Audits cannot and will not substitute for inspection, although the development and maintenance of a quality system will reduce the need for inspection by controlling production and service processes. Audits are meant to verify and measure the level of control that an organization has over its processes at the time of the audit. To provide customers with products or services that always meet or exceed requirements, organizations must continuously inspect and measure the suitability of processes and deliverables. In organizations with established quality systems, this inspection process is generally known as *monitoring*. Monitoring provides the necessary assurance that processes are being complied with every time and that nonconforming products or services are not shipped to the customers.

> ◆ Audits verify the requirements of, and controls for, the processes
> ◆ Inspections verify that the product or service meets the specification for the product or service
> ◆ Audits are snapshot views of the level of conformance to the requirements
> ◆ Inspection is an on-going activity

Chapter 2

WHY ORGANIZATIONS ARE AUDITED

Audits, whether internally or externally driven, serve as a tool to verify that an organization is actually doing what it claims. Managers and employees must understand the differences between the focus of external and internal audits in order to understand why external organizations and the organization itself are both interested in the level of conformance to the stated requirements.

The External Audit

There are two major reasons an organization might be audited by an external organization. The first is to meet the requirements of the organization conducting the audit. The second is market-driven wherein the organization itself chooses to become certified by an external organization to a particular standard, thereby enhancing market position (such as done by the U.S. automobile industry). Another possible reason would be related to specific industry-wide requirements. In this chapter we look at these two major drivers to the external audit separately.

+--+
| **External Audit Drivers** |
| |
| ◆ External Requirements |
| • Contractual |
| • Regulatory |
| ◆ Internal Requirements |
| • Enhance market position |
| • Quality system certification |
+--+

External Audits Driven by External Requirements

As a condition of doing business, many organizations will audit those with which they do business to ensure that a particular set of requirements that they have defined is being met. These audit requirements most often are spelled out in contract clauses or mandated by a regulatory body, such as a government agency, international regulatory organization, etc. Below is a brief discussion of these audit requirements and how they relate to the auditee.

Contractually Driven Audits

In all contractual relationships there is a purchaser and a supplier. The supplier has a contractual obligation to meet specific requirements, either stated or implied by the contract. The purchaser may state in the contract that all product- or service-related capabilities, facilities, and processes are subject to audits. These audits provide the purchaser the assurance that the supplier's facilities and processes are controlled. For the most part, these audits are designed to focus only on the processes involved in supplying the deliverable to the purchaser and verify that all of the requirements of the contract are being met. The contractually driven audit is most prevalent in the defense, space, and health industries, although many other industries are beginning to use audits prior to delivery to ensure that what they contract for will actually be delivered. In general, the scope of the contractually driven audit is limited to the products or services that the supplier provides to the purchaser.

Regulatory Audits

With an ever-increasing public concern for product safety, the audit has become a way to ensure that regulations are met by organizations that provide products and services. The focus of regulatory audits is the verification of conformance to written and publicly available requirements that govern the production of goods and the delivery of services. There are agencies responsible for regulating each type of industry. In the United States, these regulatory agencies include the Environmental Protection Agency, the Food and Drug Administration, the Department of Education, and the Occupational Safety and Health Administration. Each country, state, province, and city around the world has similar agencies regulating the production and delivery of goods and services. Rest assured, during the next decade an increasing number of agencies will be auditing private sector manufacturers and service providers.

Audits Driven by Government Regulations

- ◆ Focus on product safety and service delivery
- ◆ Conformance to published regulations
- ◆ Usually mandated by legislation
- ◆ Represent the public's concerns

External Audits Driven by Internal Requirements

Competition for market share spurs organizations to find ways to set themselves apart from their competitors. A common marketing approach is to claim that a product will go further toward meeting and exceeding customers' needs and expectations than a competitor's product.

Audits as a Marketing Tool

By opening up the organization's method of conducting business to the scrutiny of external auditors, the organization can advertise its success to differentiate itself from the competition.

Following the lead set by Japanese companies in the 1950s, companies worldwide are adopting quality systems as the best approach to achieving and demonstrating consistent quality, whose definition is rooted in the customer's needs and expectations.

With the establishment of quality systems comes the requirement to be audited. Initially, audits are handled internally and are used to improve the quality system. As a quality system matures and an organization gains control over its processes, the organization seeks formal recognition of its accomplishment. That recognition amounts to certification by an accredited external auditing body. A company's certification to a recognized quality standard, such as International Organization for Standardization (ISO) 9000 series, can be a powerful marketing tool.

Some standards organizations, like ISO, do not perform audits themselves, but accredit others to audit on their behalf. These accredited auditing bodies grant certification only after thoroughly reviewing a company's documentation to ensure a quality system has been established and maintained in compliance with the relevant clauses of the standard. Auditors rigorously compare what an organization's quality system documentation claims with what is actually being done. Upon completion of initial audits and subsequent approval of corrective actions, the accredited body certifies the organization.

To maintain certification, standards organizations require that periodic audits be carried out to ensure the company's quality system continues to be maintained and followed. Therefore, with the decision to become certified to a standard such as the ISO 9000 series, an organization should expect to be audited regularly by an external organization.

Companies that aspire to external quality system certification are rightfully proud, and welcome the external auditor with open arms. These companies and everyone in them have the vision to make their organizations a model for others to follow and a benchmark against which to measure themselves.

Sometimes the decision to attain external certification is not solely driven by a company's desire to be a leader in its field. The need for certification sometimes comes from a wish to do business in a particular geographical region or with a particular organization that requires certification as a prerequisite for doing business. This is fast becoming a fact of life in the European Community.

The Internal Audit

At a rapidly accelerating pace, organizations are establishing quality systems that definitively spell out requirements for all phases in the production of goods and services for the marketplace. As part of their quality systems, these organizations must establish a means by which they can verify that requirements are understood and complied with all the time. The *internal audit* is the method companies use to verify conformance to requirements and the effectiveness of the quality system. Company employees who have been trained in quality system auditing techniques conduct internal audits.

The Internal Audit

◆ Drivers
- External Requirement
  · Part of certification to external standard such as the ISO 9000 Series
- Internal Requirements
  · Management Tool
  · Quality system improvement support
  · Implementation verification
  · Awareness and training of employees

During the early stages of quality system implementation, employees struggle to understand the "new" requirements and their roles in the system. An internal audit of a quality system at the implementation stage promotes awareness and provides the company with the necessary assistance to make the system operate effectively and successfully. These initial internal audits are *prescriptive* in nature, meaning they allow the auditor to help and to act as a facilitator.

As a quality system matures, internal audits help uncover opportunities to improve both the quality system and the way the organization does business. At this stage, it is critical for an organization to have a structured way of continually measuring how well the quality system is being implemented and maintained. The internal audit process provides this measurement, as well as a solid basis for future comparisons.

Very often, organizations aspire to become certified to an external standard like the ISO 9000 series, which requires internal audits to be performed and, without which, they cannot be certified to the ISO Standard.

The ISO 9000 series is not alone in demanding adherence to quality standards. Many businesses worldwide now require that their suppliers demonstrate conformance to an internal quality management system as a condition of doing business with them.

Internal audits are used as a management tool to determine the level of quality system implementation and conformance and to identify opportunities for improvement. Through internal audits, managers and employees alike can answer many questions, helping the organization maintain its market position. Of the questions an internal audit can help answer, the following are the most critical to the health of an organization.

Internal Audits Answer the Questions

- Are we doing what we say we do?
- Are we doing it the way we said we would?
- Can we actually do things the way we say we do them?
- Do we have a weakness or potential weakness that can be corrected or eliminated?
- Can what we do be done better?

Answering these questions through the internal audit process will provide the management and employees with a common focus and vision for the future.

Regardless of an organization's reason for establishing and maintaining a quality system, the audit process is an invaluable tool for verification.

Audits as a Tool for Verifying

- That an organization conforms to the stated requirements of its quality system and other standards
- That the quality system is understood by all employees
- That the quality system has been implemented and is maintained at all levels within the organization
- That the quality system meets the needs of, and is responsive to
  - the customer
  - the organization
  - and other changing standards
- That the quality system provides effective controls to ensure conforming and consistent product and service delivery

---

Chapter 3

TYPES OF AUDITS

There are as many types of internal and external audits as there are reasons for performing them. The type of audit performed will be based on the objectives of the audit. The objective may be to measure the quality system as a whole or just a particular element of the system. The audit could be designed to examine a particular procedure or process, or system conformance to a set of requirements. The audit may focus only on a particular product or service the company provides, or a particular function or operating groups. Finally, the audit may simply be conducted as a follow-up audit to confirm corrective and preventive actions prescribed by a previous audit.

Audits Types

The objectives of the audit will determine the selection of the particular type of audit to be conducted. Each audit type approaches conformance verification from a slightly different angle, but all have the same goal. Below is a discussion of the various audit types an auditor may select.

Types of Audits an Organization May Undergo
◆ System audit
◆ Procedure audit
◆ Process audit
◆ Product or supplier audit
◆ Function or department audit
◆ Follow-up audit

<u>System Audit</u> - A system audit determines whether or not the entire quality system or some of its elements have been developed, implemented, and maintained in conformance with established requirements or an external standard.

<u>Procedure Audit</u> - A procedure audit verifies that a particular procedure or procedures meet the requirements of the system, and that work associated with the procedure(s) conforms to the requirements stated in the procedures.

<u>Process Audit</u> - A process audit verifies that a process conforms to established requirements and generally examines a process from beginning to end. The scope of a process audit is generally much narrower than that of a system audit.

<u>Product Audit</u> - A product audit focuses on a particular product or service provided by a company. The product audit evaluates whether or not product or service requirements have been met from the start of processing through process completion. This type of audit is also known as a *supplier audit*, because suppliers are often audited by their customers to evaluate whether development and manufacturing processes for a specific product are under control. Technical expertise may be necessary to conduct a product audit.

<u>Function or Department Audit</u> - A function audit is utilized to determine whether a function or department conforms to the requirements for the processes for which they are responsible. This type of audit is the most common for small companies or offices within a company.

<u>Follow-up Audits</u> - A follow-up audit verifies that corrective and preventive actions have been effective in preventing recurrence of nonconformances uncovered during a previous audit. These follow-up audits are generally short and directed only at verifying resolution of corrective and preventive actions.

Prescriptive and Compliance Audit Methods

All of these audit types may be conducted using either of the following auditing methods:

- ◆ The prescriptive audit method
- ◆ The compliance audit method

The method employed is generally based on the needs and objectives of the organization conducting the audit and the goals of the audit. The maturity of the quality system and who is performing the audit are often the deciding factor in choosing a method.

The Prescriptive Audit Method

In addition to its investigative value, the prescriptive audit is a valuable teaching tool. The prescriptive audit method is used almost exclusively during internal audits. On occasions during an external audit, particularly during a supplier audit, the prescriptive method may be used to help a supplier find the best way to meet compliance requirements, but this situation is rare.

The Prescriptive Audit Method

◆ A teaching tool:
  • Increases auditee awareness of the quality system requirements
◆ A tool for implementation
◆ A measure of how the quality system is performing
◆ A method of sharing successful methods of meeting the requirements with auditees

---

Used during the implementation phase of a quality system, the prescriptive audit method reaches three goals: 1) it allows auditees to increase their awareness of quality system requirements, 2) it assists in the implementation of the quality system, and 3) it measures how well the quality system is performing. During a prescriptive audit, the auditee plays the role of student, attempting to learn as much as possible from the auditor. Although measuring how well a quality system has been implemented and maintained is important, the greater focus in prescriptive audits is on teaching and learning. Why?

In the early stages of a quality system, the auditee may have a poor understanding or low awareness of quality system requirements. The auditee may, for instance, have well-written procedures that show how responsibility is shared and what standards are expected, but then fail to write process or work instructions, which show the step-by-step controls in a task. An experienced auditor using the prescriptive method can facilitate a better understanding of how a quality system can be implemented and improved than a manager could gain alone. The auditor brings a wealth of information about proven methods to the audit process. The auditee should take this opportunity to learn as much as possible from the auditor so that the auditor's shared knowledge can be put into action.

The Compliance Audit Method

In all types of audits, the audit is a fact-finding mission to discover whether the organization complies with the requirements it has established for providing products and services. In contrast to a prescriptive audit, a compliance internal audit is best defined as a strictly fact-finding mission to verify conformance to stated requirements and standards.

In a fully mature quality system, the compliance audit method is used almost exclusively, for both internal and external audits.

Preventive Focus

For the manager, the central objective of both internal and external audits is *prevention*. Quality system audits allow organizations to identify opportunities to improve both processes and the quality system before problems become serious enough to affect the product or service provided. The audit is one of the most powerful tools an organization has to uncover these improvement opportunities.

The audit is an effective way to help the organization improve. That is its purpose, and that purpose must be communicated clearly by management to everyone in the organization. The message that must be understood by management and employees alike is "Audits take place to help us improve and keep our processes under control." By clearly focusing on prevention, the organization fosters an atmosphere of continuous improvement and creative change in the organization.

The Preventive Focus

- The central theme in all audits
- Managers must view audit findings as opportunities to improve and prevent recurrence of nonconformances
- It is crucial to find and correct problems before they become evident to the customer

---

Chapter 4

UNDERSTANDING AND MEETING THE REQUIREMENTS

The most important action managers and staff can take before an audit is to prepare properly. This preparation cannot be an exercise that happens the week before the audit team arrives. With the appropriate preparation, the auditee can speak confidently with the auditor, anticipate where an auditor's line of questioning is going, and guide the auditor in the direction the manager wants to go.

To understand and meet the requirements, you will need:

- An attitude of improvement and cooperation

- Educating staff about the purpose of audits

- Identifying quality system requirements

- Evidence of conformance

- Self-assessment and related checklists

- Personnel training

    - Indoctrination training

    - Process specific training

    - Evidence of training

An Attitude of Improvement and Cooperation

Management's commitment to the audit is critical to the success of the organization and the audit. The staff's perception is greatly influenced by the manner in which the audit is announced. If management sets a tone that says "This audit is important. It will let us know where we are, how well we are doing, and where we can improve our products and services", the staff's attitude will more likely be positive and management will see a higher level of buy-in to the quality system and audit process. Improvement and cooperation must be the manager's attitude towards audits and requirements. In other words, management must set the tone of the audit.

If, on the other hand, management projects an attitude that says, "We have to do everything we can to hide what we do wrong," management will effectively be telling employees that the audit and the quality system are simply what we do when the auditors are there and that the quality system is not to be taken seriously. The employees will perceive the audit as a tool by which management will find guilt and lay blame. In such situations, everyone from department manager on down will be driven by fear rather than a desire to improve.

Understanding and meeting the requirements will let us know where we are, how well we are doing, and where we can improve our products and services.

Educating Staff About the Purpose of Audits

The quality system and the audit process require the cooperation of all employees. Employee acceptance of the audit process is critical to the success of the audit. To foster the desired attitude, the entire staff needs a complete understanding of the *what*, the *why*, the *how*, and the *when* of the audit. The only way to fulfill this managerial responsibility is to educate employees about what happens during the audit and how the department will use the results to improve the way products and services are developed and delivered. To develop the full support of the staff, managers must share information with employees. As a rule, when employees understand their audit roles and the importance of the auditing process, they are more willing to work as a team for the unified goal of continuous improvement.

Identifying Quality System Requirements

When preparing for a first audit, managers and employees are often unsure which requirements apply specifically to the work they perform. The quality system requirements that apply to one department may differ greatly from the requirements of another department. In addition to being aware of the relevant requirements of the quality system, managers must be aware of the industry standards that affect the overall company and the department in particular.

Managers, together with their staff, should identify the requirements through the review of:

> ◆ Procedures
> ◆ Process or work instructions
> ◆ Contracts
> ◆ Standards and regulations

Including the staff in this process will:

> ◆ Develop an understanding of the requirements among all staff members
> ◆ Rally the support of the staff in the acceptance of change and the audit process

This is not an easy task for most managers. However, during development of the quality system, most, if not all, relevant clauses of governing standards should have been identified and included in overall company policy and quality system procedures. For example, during quality system development in compliance with the ISO 9001 standard, all twenty 9001 clauses are studied. How each clause applies is then addressed in relevant quality system documentation.

Although the quality system documentation covers all elements for the overall organization, all elements do not necessarily apply to all departments. Additionally, the overall system cannot reflect all standards and contractual requirements that govern a single, specific department. It is generally considered the department's responsibility to develop procedures and process instructions that address local requirements like these. Therefore, managers and staff must review the overall quality system documentation to determine which elements apply, and review all contracts to determine which elements need to be covered in the department's procedures and process instructions. The benefit of involving staff members in the process is twofold: everyone gets a better understanding of what must be done to meet system requirements, and everyone supports the changes because they know why the requirements exist.

Once applicable quality system, industry, and contractual standards are on hand, managers and staff should highlight the specific requirements of each document, and develop checklists for use during *self-assessment*.

By identifying what applies to their respective department, managers and staff can focus on improving where they feel they have weaknesses.

Evidence of Conformance

During review of each quality system requirement, the question "Where is the objective evidence that verifies our compliance?" must be answered. Each staff member should list the type and location of documents that provide objective evidence. This exercise will uncover areas where the department is either not complying with a requirement or where it has no objective evidence to demonstrate compliance.

Once compliance documents have been identified, the department manager and staff should assess the appropriateness, accuracy, and completeness of each document. This assessment can be used to improve documentation and to ensure that a mechanism exists by which compliance with every requirement can be demonstrated.

Refer to this list when an auditor requests objective evidence:

> ◆ What constitutes objective evidence?
> - Documentation that what the organization says it does is actually done
> ◆ The manager must ask "What objective evidence do we have to show that we meet this requirement?"
> ◆ Before an audit, the department should ensure the appropriateness, accuracy, and completeness of the objective evidence

Self-Assessment

Self-assessment is a comparison of established practices to documented standards and requirements. Often, the self-assessment process is looked upon by managers and employees as a dry run for an audit. Usually, procedures and process instructions are already in place at this stage of quality system development. At early stages of quality system development, however (and at other stages, for that matter), procedures and instructions may not be routinely followed. Hence, requirements may not be met all the time. The self-assessment provides a measurement that can be used by the manager and the staff for improving the level of implementation prior to an internal or external audit.

> *Definition of Self-Assessment – The process of comparing the established practices of the department to the documented requirements, contracts, and standards*

Self-Assessment Checklists

Self-assessment checklists may be arranged by hierarchical level within the quality system, then by function, then by process. Here is the rationale behind developing checklists this way.

At the highest level of a quality system are a company's quality system manual and quality system procedures. Both types of documents set common standards and requirements for quality system documentation, corrective action, document control, quality record maintenance, and confidentiality. Since quality system procedures apply to the administration of all functions and processes, they make a good start for any self-assessment checklist.

Self-Assessment Checklists

- Address the requirements of the quality system as they relate to the department
- The requirements of the following are considered in the development of a checklist:
  - Quality system manual
  - Organization-wide procedures
  - Process related work instructions
  - Local procedures and work instructions
  - Contractual requirements
  - Standards and regulations

Going a step further, it helps to subdivide checklists for large departments into distinct functional lists. For example, the Office Services Department may perform several functions, including packing, handling and delivery, purchasing, maintenance service contracting, and storage. Meanwhile, the Customer Service or Design Department may be responsible for contract review, development and distribution of work or process instructions, service delivery, activity monitoring work control, and service or design policy. This is why it is critical in the self-assessment process that the checklists be developed by the department manager and employees during the review of the requirements.

Once the checklists have been developed to address the requirements in the quality system procedures and process instructions and the department's operating procedures and process instructions, the manager, along with the staff, can determine whether the department meets each of the requirements.

During self-assessment, the availability of objective evidence to each requirement must be noted. Objective evidence is obtained through documents and records that demonstrate requirements have been met. The most common format of objective evidence is the check sheet, which shows that each step in a process was completed and attested to by the individual carrying out the process. There may be documents other than check sheets that also demonstrate completion of each process step. Only the manager along with staff can identify these.

During self-assessment, the manager should note areas where the requirements have not been met every time or where there is a lack of objective evidence demonstrating compliance. These identified areas should be viewed as opportunities to improve the system and the manner in which the department carries out its work.

The major benefit of conducting a self-assessment is that the department will have the opportunity to identify nonconformances before an auditor uncovers them. By uncovering these nonconformances before the auditor does, the department can take corrective action without the additional pressure of having to succeed in an audit.

Self-assessment needs to be carried out weeks or even months before the scheduled audit so there is sufficient time to implement the improvements.

Benefits of Self-Assessment

♦ The major benefits of conducting a self-assessment are:
  • Provides a casual preview of the audit process
  • Allows the department to find nonconformances and observations before the auditor finds them
  • Provides a measure of the level of implementation and maintenance of the quality system within the department

---

A sample self-assessment checklist is given in Part 3–Reference Material. This self-assessment checklist has proven indispensable to department managers in preparing for internal and external audits of the company from which it was taken. This checklist should be customized to fit the respective company for which it is applied.

Personnel Training

To understand the requirements of the quality system and the work processes, every employee must be trained. Having the employees trained in the requirements will make the audit process run more smoothly and will provide the employees with the confidence necessary to be relaxed during the audit process.

Personnel Training

- ◆ A basic tenant of quality is that all personnel are trained in the processes they carry out
- ◆ The manager should assure that training has been provided for all employees in the following areas:
  - Awareness of the quality system
  - Procedures and processes that the employee carries out
- ◆ The department must be able to demonstrate that training has been completed
  - Includes all staff in the audit process

Quality System Indoctrination Training

Employee quality system indoctrination training is critical. This type of training ensures that each employee understands the requirements described in a company's quality system. Indoctrination training may be conducted company-wide or within each department.

Procedure and Process-Specific Training

Procedure-specific and process-specific training requirements may be prescribed in functional procedures and process instructions, or prescribed in other documents, such as job descriptions, to which procedures and processes refer. For example, a procedure or process instruction for ship structural plan review will state the training and knowledge required of engineers who review tanker designs. If training requirements are not found in the procedure itself, the procedure will refer to another document that lists training and knowledge requirements in detail. Since most quality systems require that only trained individuals carry out work that effect the quality of the goods or services, the auditors will investigate the qualifications and training of the individuals involved in a process.

Providing Documented Evidence of Training

Training is required for any employee who serves a function or performs a task, unless that function or task is always carried out under the direct supervision of someone else with the required training.

Auditors will be looking for detailed documentation of training. This evidence may be a certificate of course or seminar completion, relevant diplomas, or records of process-specific on-the-job training conducted by someone qualified in the process. Auditors will also be looking at the level of organization of a department's training documentation. Auditors would like to see that a department has the ability, at a glance, to determine its training needs for the near future. Training records and training schedules should be brought up-to-date before the audit. Additionally during the preparation of an audit, the department manager and staff should review new and newly revised standards, procedures, and process instructions to identify new training requirements that might have been overlooked.

Sample Self-Assessment Checklist

Element or Requirement	(Circle one)		
Is file management maintained?	No	Yes	
Is correspondence requiring action tracked?	No	Yes	
Is the corrective action system maintained?	No	Yes	
How is the general housekeeping?	Needs work	Almost there	In good shape
Are controlled documents up-to-date? (procedures, PIs, circulars, etc.)	No	Yes	
Are quality records generated and maintained?	Some of the time	Most of the time	All of the time
Are back-up requirements identified and maintained?	Some of the time	Most of the time	All of the time
Are all transmittal records returned?	No	Yes	
Is sub-issue of controlled documents done in a controlled manner with transmittal evidence?	No	Yes	Not Applicable
Are local procedures & process instructions approved?	No	Yes	Not Applicable
Are local procedures & PIs distributed and maintained in a controlled manner?	No	Yes	Not Applicable
Are there any uncontrolled copies in the department?	No	Yes	
Have all blank forms and check sheets which have been superseded been removed from the files and workplace?	No	Yes	Not Applicable
Have all local controlled documents been numbered as required?	No	Yes	
Have all quality records been numbered as required?	No	Yes	Not Applicable

Element or Requirement	(Circle one)		
Have personnel forecasts been kept up-to-date?	No	Yes	
Have the training needs assessments been done?	No	Yes	
Do all positions have position descriptions?	No	Yes	
Are all personnel aware of their position descriptions?	No	Yes	
Is a departmental training plan available and is it kept up-to-date?	No	Yes	Not Applicable
Have employees submitted training update forms to HRD for completed training courses?	No	Yes	Not Applicable
Are all employees certified in the processes they carry out?	No	Yes	
Has the training certification been entered in the training database?	No	Yes	
Have CARs been generated, resolved and closed in accordance with applicable, procedures?	Some of the time	Most of the time	All of the time
Have follow-up action plans been documented and completed from the last internal audit?	No	Yes	
Have local and worldwide procedures been developed when needed in accordance with the applicable requirements?	Some of the time	Most of the time	All of the time
Is there evidence of the required reviews and approvals for procedures?	Some of the time	Most of the time	All of the time
Have editorial revisions and substantive revisions when recognized been brought to the attention of those who can make the connections?	Some of the time	Most of the time	All of the time

Element or Requirement	(Circle one)		
Has the department recruited or hired anyone in the past year?	No	Yes	
If yes, has this been done in accordance with company procedures?	Some of the time	Most of the time	All of the time
Have the following been documented for projects: scope of work, project management, project tracking, project schedule, progress meetings and kick-off meetings in accordance with company/department procedures?	Some of the time	Most of the time	All of the time
Do the position descriptions include: title, general summary of work performed, principal duties and responsibilities, training and knowledge requirements disclaimer clause, and reporting relationships?	Some of the time	Most of the time	All of the time
Is a current master list of company owned publications available?	No	Yes	
Has the department been involved in any of the reviews of requests for services?	No	Yes	
If yes, have all clauses relevant to the review and processing been complied with?	Some of the time	Most of the time	All of the time
Is every project uniquely identified with a project number, batch number or similar information?	Some of the time	Most of the time	All of the time
Is all project related documentation traceable to the project?	Some of the time	Most of the time	All of the time
Do projects have identified milestones which consider the scope of work and customer requirements?	Some of the time	Most of the time	All of the time

Element or Requirement	(Circle one)		
Is the company confidentiality policy stated and understood by everyone?	No	Yes	
Is the confidentiality followed?	Some of the time	Most of the time	All of the time
Has unsolicited client feedback been received within the office/department?	No	Yes	
If yes, have records of client feedback progressed through the company in accordance with company requirements?	Some of the time	Most of the time	All of the time
Does all generated correspondence contain the elements required to effectively identify, file and track it?	Some of the time	Most of the time	All of the time
Are there correspondence file reference guidelines readily available within the office/department?	No	Yes	
Does the department use any sub-contracted technical personnel?	No	Yes	
If yes, have the selection, contracting, and work control provisions of relevant procedures been followed and documented?	Some of the time	Most of the time	All of the time
Does a locally controlled approved supplier list exist which lists the approved sub-contractors?	No	Yes	
Has the list been reviewed to determine whether any sub-contractor should be removed? Is this review documented?	No	Yes	
Have local file management procedures been developed for the department? Are they necessary?	No	Yes	

Element or Requirement	(Circle one)		
If yes, do they meet the company requirements?	No	Yes	
Have the required check sheets for the processes carried out by the department been completed in accordance with documented requirements?	Some of the time	Most of the time	All of the time
Have all suppliers of critical products or services been assessed?	No	Yes	
Do records of the assessment to defined requirements exist?	Some of the time	Most of the time	All of the time
Does a locally controlled approved supplier list exist listing the approved critical suppliers?	No	Yes	
Has the list been reviewed to determine whether any supplier should be removed? Is this review documented?	No	Yes	
Have all other provisions of the assessment and contracting of critical suppliers been complied with?	Some. of the time	Most of the time	All of the time

Chapter 5

SETTING THE STAGE

The first minutes of an audit make a lasting impression. Two ways to make a good first impression are by demonstrating a high level of organization within the department, and by making sure that logistic wrinkles have been smoothed out before the auditor arrives. This chapter covers audit preparations that are outside the realm of quality system conformance, but equally important to the smooth flow and success of the audit.

> **Setting the Stage**
> - Organizational charts
>   - Conventional versus Ishikawa charts
> - Manuals, procedures, and process instructions
> - Logistics
> - Housekeeping
> - Availability of staff
> - Audit escorts

Organizational Charts

To plan and carry out an effective audit, an auditor needs to understand the overall structure of the audited functions. While conventional organizational charts generally show who works for whom, they are not specific about the functions performed within each group.

An Ishikawa chart gives a clearer picture of the functional responsibilities within an organization. The Ishikawa, or *fishbone*, chart graphically shows the functions performed within each group. Figure 5-1 shows a typical reporting-line chart and Figure 5-2 shows an

Ishikawa chart. As is evident, the detailed fishbone chart allows an auditor to prepare a more meaningful schedule of audits based on related functions.

Conventional Hierarchical Charts Versus Ishikawa or Fishbone Diagrams

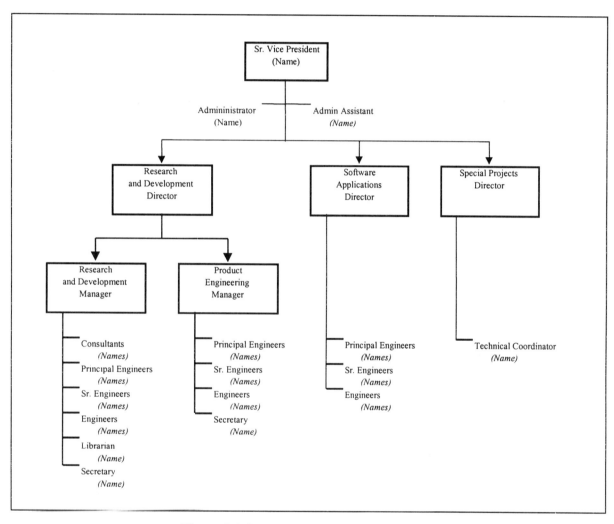

Figure 5-1 Conventional Hierarchical Chart

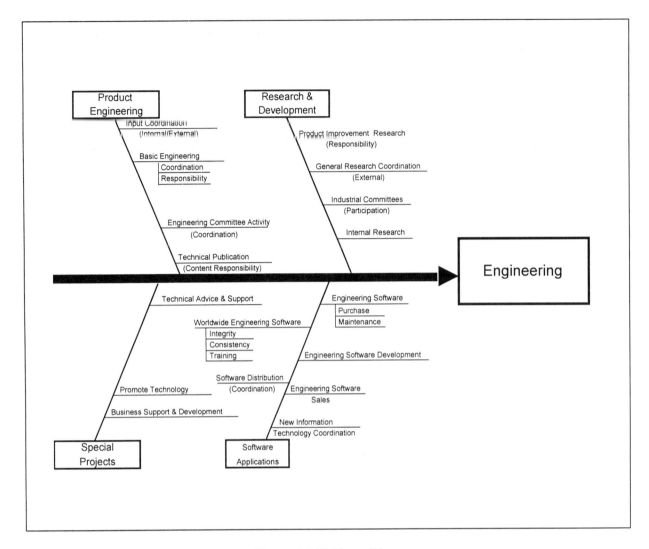

Figure 5-2 Fishbone Diagram

Manuals, Procedures, and Process Instructions

The auditee is responsible for providing copies of any supporting documentation relating to the processes being audited. These documents include the company quality system manual, quality system and functional procedures, process or work instructions, and any contracts that impose production and delivery requirements.

If any of these are controlled documents, be sure copies sent to the auditor are clearly labeled UNCONTROLLED COPY. The procedure and process instruction versions sent to the auditor for planning purposes should be current, and should reflect the way work is actually carried out, since these are the documents to which the function will be audited.

Documents Provided to the Auditors

- ◆ Quality system manual
- ◆ Functional and quality system procedures
- ◆ Work or process instructions
- ◆ Other documents that establish requirements

Logistical Considerations

Some auditors may require assistance to arrange local living accommodations and transportation. An offer of assistance should be conveyed during the audit planning stage. This may include both living accommodations as well as transportation.

At the same time, the audited company should assure the auditors of the availability of clerical services, quiet office space, telephone and facsimile services, and adequate space in which to conduct pre- and post-audit meetings. Remember: the auditors are your guests; show concern.

Housekeeping

Although auditors verify conformance to stated requirements based solely on written evidence, auditors are people like everyone else. The physical organization and tidiness of the facility itself will influence the auditor's perception. An auditor who senses disarray may decide to dig deeper and demand more written evidence to rule out doubt. Therefore, it is a good practice to devote time to the appearance of the facility before the auditor arrives. Everyone understands the importance of good first impressions. A clean, well-organized plant or office will reflect well on the effectiveness of the controls the organization has over its processes.

The auditee will reap another benefit of having a well-organized facility. If clearly labeled files and books are easily retrieved, the auditee will avoid wasting the auditor's time and patience, and give the impression that processes are under control.

Availability of Staff During the Audit

To verify that requirements are generally understood, an auditor will want to talk with a broad cross-section of employees involved in any process. Since employees who perform a process are generally also the most knowledgeable about it, managers should ensure that the key personnel that the auditor needs to speak to are available during the audit.

When a process flows across department interfaces, coordination will be required to ensure that personnel in several departments are available when needed.

Audit Escorts

In addition to scheduling staff who will be directly audited, at least one employee should be designated as *audit escort*. An audit escort should be thoroughly familiar with the quality system, the process requirements and the personnel being audited.

The escort serves several functions during an audit:

◆ Acts as an advocate to the auditee by noting the auditor's reactions, by providing a "second listener's" interpretation of an audit question

◆ Acts as a note-taker, allowing the auditee to speak with the auditor without being distracted by having to take notes

◆ Helps clear up any confusion the auditee or auditor may have

Chapter 6

CONDUCTING THE AUDIT

While the managers are preparing for the internal or external audit, so are the auditors. The auditors will be reviewing the quality system documentation, developing checklists, and forming a strategy for conducting the audit. The audit itself will be very much like the self-assessment, although the individuals verifying conformance with the system will be independent of the function being audited. If the auditee and the auditor have done their preparation well, the audit process should run rather smoothly.

Both internal and external audits are conducted in accordance with specific documented requirements of procedures and process instructions. Sample procedures and process instructions for *internal audits* are presented in Part 3–Reference Material. An *external* auditor's procedures and process instructions will be similar, although they may contain additional requirements mandated by the certifying body. The auditee may request copies of an external auditor's procedures, guidelines, and process instructions prior to commencement of the audit.

Auditing, both internal and external, is a process that consists of:

> - The opening meeting
> - The audit investigation (interviews and document examination)
> - Daily audit debriefings – optional
> - The closing meeting

Before the audit, the auditor and auditee will jointly establish the date, time, and location of the audit. The audit schedule will be sent to the auditee prior to the audit date. The amount of lead-time for notification and submission of the schedule is usually specified in the auditor's procedures. The two parties should reconfirm the schedule and all other arrangements a few days before the audit so that the audit can begin on schedule.

Opening Meeting

Each audit begins with an opening meeting. This meeting is run by the auditors, and is meant to familiarize the auditee with the audit process as implemented by the particular auditors.

The opening meeting agenda will normally be as follows:

1. The auditors will introduce themselves.
2. Any special qualifications of the auditors will be made known.
3. The purpose of the audit will be explained.
4. The plan for deploying the auditors will be made clear (that is, either individually or in teams)
5. A general description of the methods used in conducting the audit and the rationale for using the selected methods will be presented.
6. The identification process for observations and nonconformances will be described.
7. The terms *observation* and *nonconformance* will be defined. This is necessary because these definitions vary from organization to organization.
8. Ranking of nonconformances will be made clear if the auditors choose to rank nonconformances (e.g. major and minor nonconformances).
9. Escorts and their functions will be identified.
10. Any restrictions on access to facilities and information that the auditee will impose on the auditor will be made known.
11. Any logistical information that must be clarified prior to the audit investigation will be presented.
12. Any question that either party has will be addressed.

The Opening Meeting

- Is the official start of the audit
- Is run by the auditors
- Familiarizes the auditee with the audit process as implemented by the auditors
- Follows the agenda set by the auditors
- Is attended by key management personnel, department managers, audit escorts, and others at the discretion of the management

---

Many times, auditors will ask the auditee to make a brief presentation about the structure of the organization and physical layout of facilities, highlighting key players and functions at the facility being audited. Therefore, as an auditee, be prepared to present this type of general organizational information at the opening meeting.

It is management's responsibility to ensure that key staff members attend the opening meeting, because this meeting sets the stage for the entire audit. Additionally, although not mandatory, senior management and executives should demonstrate their commitment to the quality system and auditing process by attending the opening meeting. The auditors will expect that the meeting will start as scheduled and run for the allotted time as shown on the audit schedule. It is everyone's job to make sure this happens.

The Audit Investigation

At the conclusion of the opening meeting, the auditors will ask to be taken by the escorts to the first department or function. The escorts, with audit schedule in hand, will know where to begin. Upon arrival there, the escort should introduce the auditor to the auditee and allow the auditor to begin the investigation.

The auditor has been trained, through formal classroom training and through experience, to form an impression by observing general conditions. This is the point at which all the pre-audit effort to maintain good housekeeping pays off. The auditor will gain a certain confidence in the controls exhibited by an auditee if the area is tidy and looks controlled.

The auditor will begin by asking the auditee questions. Generally, the department manager is the first person the auditor interviews. The auditor will ask the department manager questions to ascertain whether or not processes are carried out in conformance with stated requirements. In general, auditors will follow their checklists very closely, and deviate only when intuition or some objective evidence suggests that a particular topic merits exploration in greater depth.

An experienced auditor is adept at asking open-ended questions (questions that cannot be answered by a simple *yes* or *no*). Auditors want an auditee to talk as much as possible to gain as much information as possible to form the basis for their conclusions. It is incumbent upon the person being interviewed to answer the auditor's questions truthfully and succinctly, providing only the information for which the auditor asks. The auditee should not embellish answers with additional information or opinions. If an auditor needs or wants more information, the auditor will ask more questions. Additionally, if the auditee feels that the information provided by the answer is not enough, the auditee may inquire whether the response has answered the question to the auditor's satisfaction. Doing this will provide the auditor with the impression that the auditee is open to questioning, not evasive, and believes in the audit process.

After questioning the department manager, the auditor may question some of the other staff members. It is easier for an auditor to verify compliance directly where processes are carried out and by directly questioning the individuals who carry out the processes. Therefore, every staff member must be prepared and available to speak with the auditors.

During the interview, remember:

> ◆ Auditors gain information to verify conformance by interviewing department managers and staff about their processes
>
> ◆ The auditors try not to ask questions that can be answered "yes" or "no"
>
> ◆ During the audit interviews the auditor wants the auditee to do most of the talking
>
> ◆ Managers should school auditees to answer questions without embellishing the answers with additional information or opinions
>
> ◆ Answer all questions truthfully

Keep the 90/10 Rule in mind:

> *Auditors have a 90 percent chance of finding the 10 percent that are not in conformance. Be honest and don't try to hide things - auditors have a way of finding them.*

Questions alone are not sufficient to verify compliance. The auditor will also review documents and records. Auditors generally use a random sampling technique when reviewing documentary evidence, and will expect to be able to choose their own samples. The auditee should allow the auditor to view any document that the auditor desires. The only exception to this would be documents that have been indicated to the auditor as being confidential in nature and to which access must be restricted. In this situation, explain the reasons to the auditor. Most auditors respect the confidentiality of the auditees documents and will seek another suitable sample to verify compliance.

> **Auditors Look for Objective Evidence**
>
> ◆ In addition to interviewing the auditee, the auditor will be looking for documented objective evidence which verifies the auditees claims
>
> ◆ The auditor will look at samples of the evidence to assure that what is claimed to be done is done every time
>
> ◆ Auditors will select their own samples

There will be occasions, no matter how well the quality system has been developed, implemented, and maintained, when the auditor will uncover either observations or nonconformances. The auditor will tell the auditee when observations or nonconformances are identified. These will be discussed with the auditee at the time of discovery. This is one opportunity the auditee has to clear a misunderstanding. If the auditor has indeed uncovered an observation or nonconformance, the auditee should acknowledge this and allow the auditor to move on with the audit. Never, as an auditee, argue with the auditor. Look upon the observations and nonconformances as opportunities to improve the quality system and how work is done.

Observations versus Nonconformances

An observation is a detected weakness, which if not corrected, may result in the degradation of service quality.

A nonconformance is a non-fulfillment of a specified requirement.

If the auditee has a real problem with an observation or nonconformance, the auditee should state that he or she feels the observation or nonconformance was raised because of a misunderstanding. The auditee will have the opportunity to present evidence demonstrating that the observation or nonconformance does not exist at the closing meeting or in the action plan that follows the audit.

When the auditor completes the investigation in one department, he or she moves on to the next area. This does not mean that the audit of the department is over. The auditor may decide to return at a later point to ask follow-up questions based on the investigation of another department that has an organizational interface with the department previously audited.

The audit investigation process will be repeated for each department or function on the auditor's schedule. Auditors will try their best to stay on schedule. Auditees and escorts should try to facilitate this to the greatest extent possible.

Daily Audit Debriefings

When audits extend over several days, auditors may wish to conduct daily audit debriefings with the auditees. These debriefings act as a series of mini-closing meetings. The department managers being audited during the course of the day and other key management personnel generally attend daily debriefing meetings. At the debriefing meetings, the auditors will inform the auditees of the nonconformances and observations uncovered during the day's investigation. The auditees will be given an opportunity to ask questions about the observations and nonconformances as well as present their positions concerning questionable observations or nonconformances raised by the auditors. At the daily debriefing, the auditors may request that a member of management, usually the management quality representative, sign off on the nonconformances. This is generally just a formality, as the auditor will, in the audit report, formally present the observations and nonconformances uncovered during the audit whether they have, or have not been, accepted. It should be understood that the daily debriefing sessions do not replace the closing meeting and primarily serve to make the time spent at the closing meeting more manageable.

Daily Audit Debriefings

◆ Auditors present a summary of the day's audit activities
◆ Observations and nonconformances are discussed
◆ Act as mini-closing meetings
◆ Daily audit debriefings are optional
◆ Attendees should be limited to key management personnel, audit escorts, and department managers whose departments were audited that day

---

The Closing Meeting

The closing meeting is the final opportunity for the auditors and auditees to get together to review the activities and results of the audit. The main objective of the closing meeting is to ensure that the auditee fully understands all observations and nonconformances identified during the audit. This meeting is generally scheduled immediately following the conclusion of the audit. All individuals who participated in the audit should attend the closing meeting.

A typical agenda for the closing meeting will include the following:

1. *Review of the purpose and scope of the audit* - This will be a brief summary of the extent of the audit, outlining the processes covered and why the audit was being conducted. During this time, an attendance register will be passed around for signing by all present.

2. *Rules of the meeting* - The lead auditor will define the format of the meeting so that it is effective and efficient. The auditor may choose to review all the findings and then have discussion, or have discussion after reviewing each finding. The auditee should abide by the rules presented by the auditor. The auditee should remember that the auditors preside over the closing meeting.

3. *Presentation of the summary of observations and nonconformances* - The lead auditor will review each finding with the auditee if not done during daily debriefings. As findings are reviewed, the auditee will be asked to acknowledge receipt of each nonconformance. This can be accomplished by simply signing the nonconformance. The auditee should accept the nonconformances and use the appeal procedure established by the auditing

organization if it is felt that the nonconformance is unjustified. Unless the auditor retracts the nonconformance immediately after discussions are held, it is unlikely to be retracted at the closing meeting.

4. *Review of auditees responsibilities* - The responsibilities of the auditee after the audit team leaves will be covered. These responsibilities include, but may not be limited to:

   a. *A follow-up action plan* – This is generally required by the auditees' procedures.

   b. *A response to the auditor by an agreed upon due date* - Follow-up plans are usually required within thirty days of receipt of the audit report.

5. *The audit report* - The auditor will inform the auditee of the target issue date of the audit report. Additionally, the auditor will generally provide the auditee with a verbal summary evaluation of the audit and advise if a follow-up audit is recommended. The summary evaluation should provide the auditee insight into the tone and message that will be conveyed in the audit report.

6. *Question and answers* - The auditee will be given the opportunity to ask any questions or clarify any issues since that will be the last time all the audit participants will be together. The closing meeting is not the place to argue with the auditors.

7. *Close* of the meeting - The auditors will close the meeting by expressing their thanks for the assistance provided during the audit. The auditee should also thank the auditors for their efforts and ask if any transportation or other arrangements need to be made.

The Closing Meeting

◆ Main objective of the closing meeting is to present:
  • Observations
  • Nonconformances
  • Summary of the audit findings
  • The auditees responsibilities for follow-up
  • Information about the audit report
◆ Auditees will be afforded their last opportunity to ask the auditors questions about the audit results.
◆ The meeting is usually attended by the same personnel who attended the opening meeting, although others may be invited.

Chapter 7

COPING WITH AUDIT FEAR

For the first-time auditee, an audit is an unknown. Initial audits are often hastily arranged directly on the heels of quality system implementation. During the quality system implementation, the auditee struggles, with visible dismay, through the writing of procedures and process instructions — often for the first time. Having done that, the auditee will attempt to work according to the procedures, wondering if everything meets quality system requirements and wondering, finally, if it's all worth the effort.

- ◆ Fear has three opportunities to build
  - • Before the audit
  - • During the audit
  - • After the audit

All auditees face audit fear — a statement that holds true for experienced auditees as well as first-timers. Fear occurs at three points in the audit process: 1) in anticipation of, 2) during, and 3) following the audit.

Pre-Audit Fear

Pre-audit fear would be easy to deal with if it had one cause, but it has many. In the first place, a quality system is an imposed order on daily activities. All professionals have, as all managers know, pride in what they do and how they do it. Thus, a certain amount of resentment accompanies imposed order. Many an employee's first reaction to the implementation of a quality system is, "I don't know why I'm doing this — I don't need to justify myself to anyone". Come audit time, that resentment may be turned toward the auditor, whom employees see as *cop*, *judge*, and *jury*, all rolled into one. Extending this logic, the role of *executioner* is reserved for the manager, and that's where *resentment becomes fear*; fear that a close examination will reveal personal weaknesses that may result in delayed promotion, demotion, or even job loss.

In the current austere business environment, employees have another fear: *job rationalization*. Documented quality systems make work processes much more visible to the employee. When all parts of a work process become visible, employees fear it will become apparent that what they do is not necessary. Fortunately, this fear is unfounded. Quality systems and the philosophy of total quality management are implemented because they pay off in the end, not because they reduce work along the way.

"Before the Audit" Fears

◆ The auditor is the cop, judge, and jury

◆ Management is the executioner

◆ The auditees feel resentment of the need to justify their methods, which becomes fear

---

The manager is to reassure the auditee that audits look at the system and processes, not the individual.

Fear During the Audit

A company's first audits will be conducted by its own employees, usually by those who are most familiar with the processes and the quality system. Many experts view the first internal audits in a positive light — as a demonstration of a company's self-empowerment. Company employees, however, may see danger signs everywhere. They fear that audit confidentiality will be compromised, and that the results of an audit may go against them personally. Additionally, an auditor may be an ex-department colleague who moved up and out of the department. If the dynamics of the changed relationship have not settled, friction or uneasiness may result.

The most common cause of fear, however, is an auditees' limited knowledge and understanding of the goals and requirements of the quality system. Those limits have followed the employee from the process writing stage through the documented-use stage. Often, the employee will have made this journey with too little critical feedback from those who know better. Now the employee faces imminent accountability for compliance, without really knowing what compliance means.

Without having been through a quality system audit, employees cannot know what appropriate audit behavior is. Some employees labor, for example, under the false assumption that auditors are impatient, wanting quick, concise, answers. Total recall is *not* what the auditor is after. True, auditors have a limited time to learn as much as possible, but documentation — *written objective evidence* — is the quality system's key to providing complete, concise answers. An auditees' oral answers, however good, are only a starting point from which the auditor begins the search to verify conformance.

"During the Audit" Fears

♦ Auditees take questioning personally

♦ Auditee fear problems will be uncovered by the auditor

♦ Auditees have a lack of understanding of expectations

It is management's responsibility to educate the auditee in the process.

Post-Audit Fear

However reassuring an auditor may have been, the auditee will experience uncertainty between the end of the audit and the audit closeout meeting. Auditees need feedback. Further, some interviewed employees, though eager for feedback, do not attend the audit closeout meeting where the auditor makes a summary report. Employees who find audits taxing will fret and worry until someone rescues them. If feedback is withheld long enough, employees may become truly soured to the audit process.

Finally, though audits are intended to prove *conformance* to the quality system, nonconformances will be found, and employees will take them personally, though they should not. The quality system audit is a test of the system, not of individuals.

Employees may also be fearful that nonconformances will be held against them personally rather than against the quality system that was audited. The employee with this fear thinks that nonconformances written during the audit will give management the impression that the department has not pulled its weight. These employees also fear that disclosure of nonconformances will result in delayed promotions, demotion, and possibly the loss of their jobs.

"After the Audit" Fears

♦ Employees fret while awaiting the results of the audit
♦ Employees fear being blamed personally
♦ Employees fear the nonconformances will have a negative effect on their careers

---

Coping With Pre-Audit Fear

Managers should begin to prepare their employees as far in advance of the audit as possible. Preparation means educating employees, repeatedly, on the objectives of the quality system and the audit process. A quality system is a system to ensure uniform production and delivery of products and services that meet customers' needs and expectations. The key word is system; the *system* is tested during an audit, not the individual. The employees must become familiar with this most basic of principles.

Management must prepare all employees well in advance of the audit:

> ◆ Teach the employees the objectives of the quality system
> ◆ Teach the employees the objectives of the audit
> ◆ Reinforce the attitude of professionalism
> ◆ Dispel fear for job security

Managers must also make sure employees understand that the first audits will be conducted internally and will be prescriptive audits. Prescriptive auditors will do everything they can, upon spotting a nonconformance, to lead the department or function to an appropriate solution. Auditors who conduct first audits know how overwhelming quality system implementation can be. Prescriptive auditors call themselves *facilitators*. Knowing this, employees will be more receptive to the help a prescriptive auditor can give.

> **Preparation - A Tool to Combat Pre-Audit Fear**
> **Part I**
>
> ◆ Teach the objectives of the quality system:
> - The system will ensure uniform production and delivery
> - The system will ensure responsiveness to customers' needs
> ◆ Teach the objectives of the audit process
> - The process will determine the level of quality system implementation
> - The process is not intended to test the individual
> - Prescriptive audits facilitate compliance

Quality system implementation is a team effort. Managers will be the first in their departments or functional groups to know a quality system audit's scope. Within a group, a manager may act as leader of an *audit readiness team* or, if a group is large, a facilitator for one or more audit readiness teams. Moving forward with audit readiness as a team will, at audit time, put the proper emphasis on teamwork rather than individual success. A team approach to self-assessment is a particularly effective method, especially when *all* employees are involved.

Managers should actively reinforce the professionalism many employees feel is threatened by "imposition" of a quality system. If managers can promote the idea that the quality system is an added dimension of professionalism, employees may be persuaded that the quality system will help them achieve higher personal standards of professionalism.

Lastly, managers need to reassure employees that the quality system and total quality management are not a threat to their jobs. Since rumors can run wild, this reassurance must be an active, ongoing campaign.

Preparation - A Tool to Conquer Pre-Audit Fear
Part 2

◆ Building the attitude of professionalism
  - Quality is an added dimension of professionalism
  - Quality is a higher personal achievement
◆ Fear for job security
  - The manager must treat this fear as an ongoing concern

Coping With Fear During the Audit

There is a risk that an auditee will take questions and comments personally. Without making employees feel like small fish in a big pond, managers should, at opportune moments, stress to employees that it is the system being scrutinized and not the employee. From a theoretical standpoint of quality audits, the relevant questions are:

- Is a quality system in place?
- Is the quality system being complied with?
- Can the quality system be complied with?
- Are instructions clear enough that a similarly skilled professional could meet the system requirements by following them?

Managers must emphasize that employees are not expected to recite procedures and processes by rote. Teach employees to understand that quality records and other relevant documents demonstrate compliance with the system, and train them to be able to retrieve these documents when asked for by the auditor. Before the audit, auditors will have reviewed many of the departments documented requirements. Auditors need to see that employees know how to find, recognize, and use the latest published standards.

Coping With Fear During the Audit

- ◆ Depersonalize the audit process by focusing on:
  - Does the standard exist?
  - Do the procedures and work instructions reflect it?
  - Are the procedures and work instructions in use?
  - Could a similarly skilled professional follow the procedures and work instructions?
- ◆ Increase real-time understanding of expectations:
  - Provide escorts who know the quality system and can put the auditee at ease
  - Have the auditor elaborate when questions exist

Managers can also pass on a few important relaxation techniques to help employees relieve stress during an audit:

- ◆ Do not work late the day before an audit.
- ◆ Get a good night's sleep the night before.
- ◆ During the audit, relax and listen attentively.
- ◆ Feel free to ask the auditor to repeat or rephrase a question.
- ◆ Don't mentally add up hits and misses — that's not the point.

Since internal audits of new quality systems are prescriptive, managers should, when introducing the auditor to an employee, clearly state: "the auditor is here to help". This introduction will give the auditor an immediate opening to elaborate on that theme and should get the employees accustomed to dealing with auditors.

Dealing With Post-Audit Fear

The key to allaying fear after the audit is *timely communication*. The manager will be the first to hear the auditor's summary report presented during the closing audit meeting. If the closing meeting is delayed significantly, that fact should be communicated to everyone involved in the audit, even those who won't be attending the closing meeting.

At the closing meeting, the auditor will attempt to present a concise, object-oriented summary of the audit. The manager should pay close attention to the words an auditor uses, for the auditor will be trying to phrase things in a non-threatening, professional way. These are the same words a manager will need to let his or her employees know the audit outcome. The auditor will attempt to phrase ideas using the *passive voice* and will avoid directing blame; for example, "In three instances, no provisions were made for safe storage of products requiring rework, although the standard required them" as opposed to "Although the standard required it, three employees did not provide storage for..." The auditor will phrase statements about conformances similarly. "In production of product A, all documented requirements of the process instruction and procedure were met."

If nonconformances are uncovered, the manager can reduce employee fear of repercussion by empowering individuals or teams to solve problems *on behalf of the department or group*. Later, results can be objectively communicated to the whole department. Keeping everyone informed will go a long way toward reducing pre-audit fear of the next audit.

Coping With Fear After the Audit

- ◆ Management should keep all employees informed
- ◆ Let the employees know about delays when necessary
- ◆ Involve all employees in the follow-up and corrective and preventive action process

Chapter 8

AUDIT TECHNIQUES

The purpose of a quality system audit is to verify that processes are carried out in conformance with prescribed requirements. However, since a thorough examination of all documentation would be impossible, just from the standpoint of time, no audit can guarantee one hundred percent conformity of any process.

Just as political pollsters have sampling methods that tell them the voting habits of the electorate based on a few samples, quality system auditors have methods that tell them how well an organization is doing in its overall quality system compliance based on representative samples. Samples are selected by the auditor at random, and may number as few as two or three or as many as twenty or more, depending upon what the auditor is trying to learn.

The sampling techniques discussed in this chapter are *trace forward, trace backward, random selection, and functional responsibility*. Based on time constraints, process complexity, and personal experience, an auditor will choose a mix of these four techniques to form a representative picture of conformance to the quality system.

Audit Techniques

◆ During an audit, the auditor may employ various audit techniques, including:
  - Trace forward
  - Trace backward
  - Random selection
  - Functional responsibility

Tracing Techniques

Tracing is an orderly way of following the flow of a process from start to finish by sampling at each stage for conformance and traceability.

- ◆ Follows the flow of the process from start to finish or vice versa
- ◆ In tracing, the auditor verifies conformance and traceability by:
  - Sampling documents
  - Questioning auditees
  - Looking at the process inputs, decision points, and outputs
- ◆ Tracing provides the auditor with the big picture view of the process
- ◆ The auditee can assist the auditor by having flow charts of the process

---

Flowcharts are the clearest way to trace the flow of most processes. Flowcharts show what inputs are used in a process, the steps, major decision points, *yes* or *no* questions, and where a process leaves one group and passes to another.

An example is shown in Figure 8-1, on page 51. This flowchart shows in detail an example of corrective action, which meets the requirements of ISO 9000.

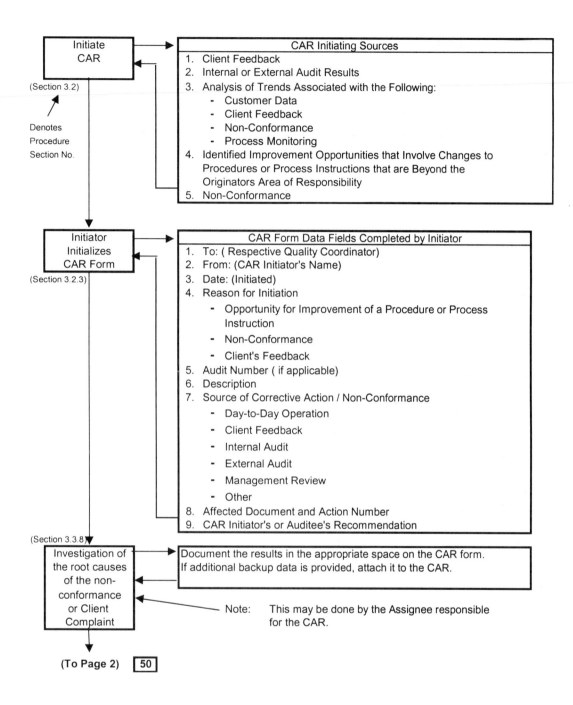

Figure 8-1 Flowchart showing example of corrective action

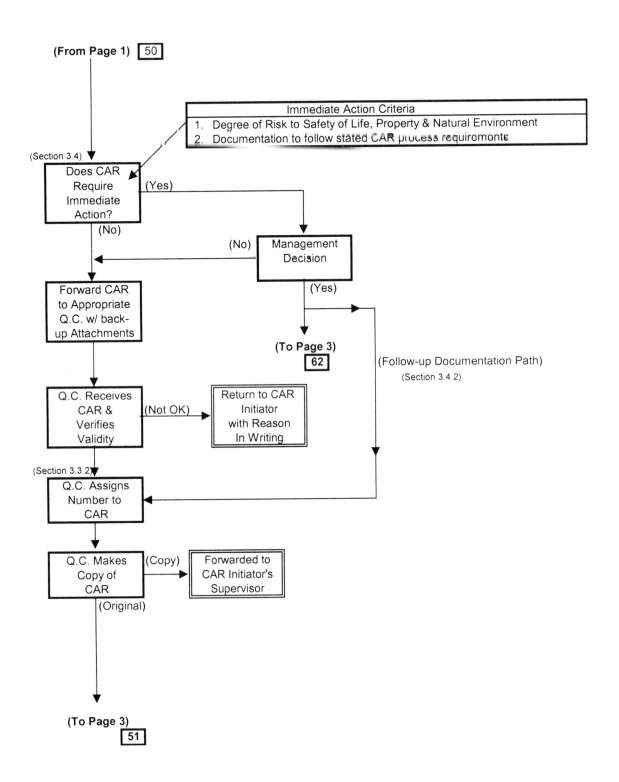

Figure 8-1 Flowchart showing example of corrective action (continued)

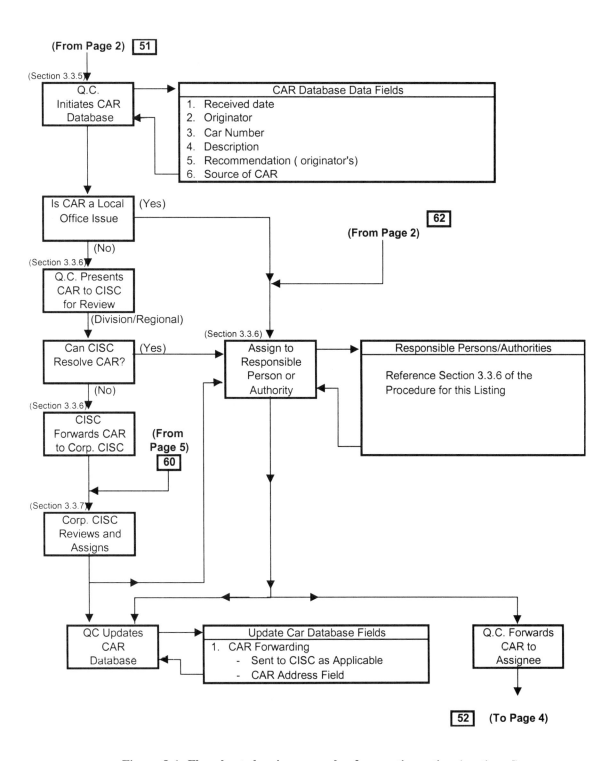

Figure 8-1 Flowchart showing example of corrective action (continued)

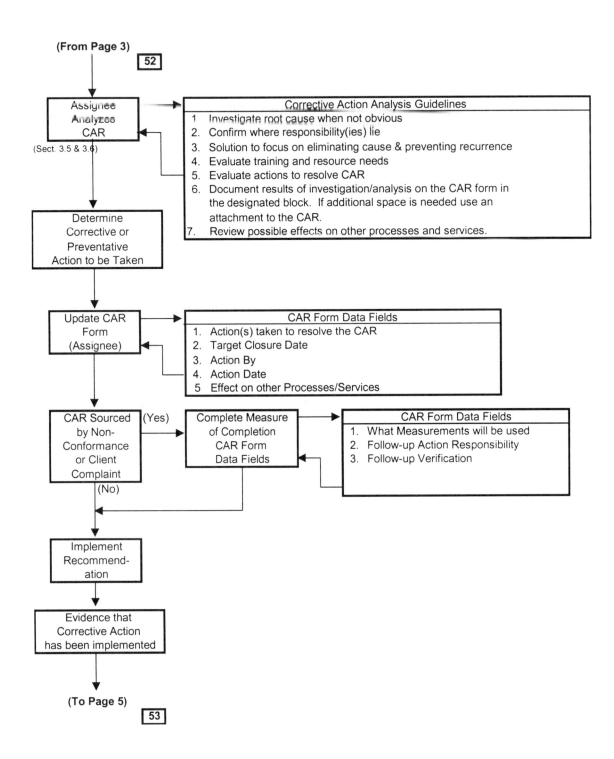

Figure 8-1 Flowchart showing example of corrective action (continued)

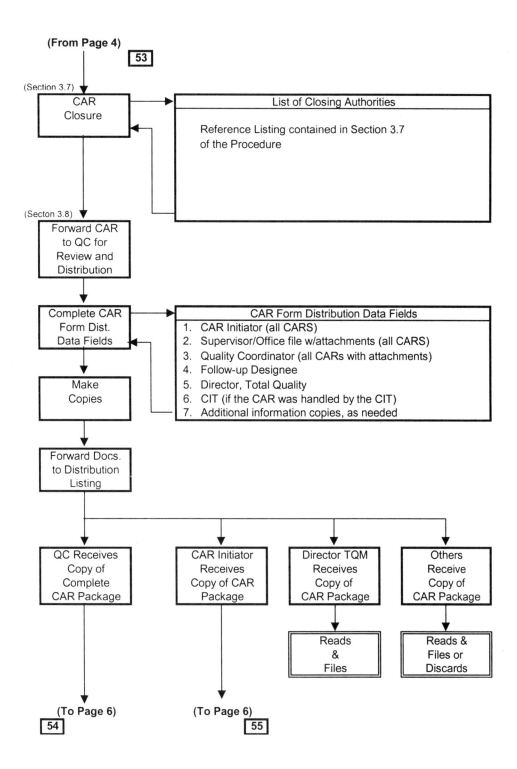

(From Page 4)

53

(Section 3.7)

CAR
Closure

List of Closing Authorities

Reference Listing contained in Section 3.7
of the Procedure

(Secton 3.8)

Forward CAR
to QC for
Review and
Distribution

Complete CAR
Form Dist.
Data Fields

CAR Form Distribution Data Fields
1. CAR Initiator (all CARS)
2. Supervisor/Office file w/attachments (all CARS)
3. Quality Coordinator (all CARs with attachments)
4. Follow-up Designee
5. Director, Total Quality
6. CIT (if the CAR was handled by the CIT)
7. Additional information copies, as needed

Make
Copies

Forward Docs.
to Distribution
Listing

| QC Receives Copy of Complete CAR Package | CAR Initiator Receives Copy of CAR Package | Director TQM Receives Copy of CAR Package | Others Receive Copy of CAR Package |

Reads
&
Files

Reads &
Files or
Discards

(To Page 6)

54

(To Page 6)

55

Figure 8-1 Flowchart showing example of corrective action (continued)

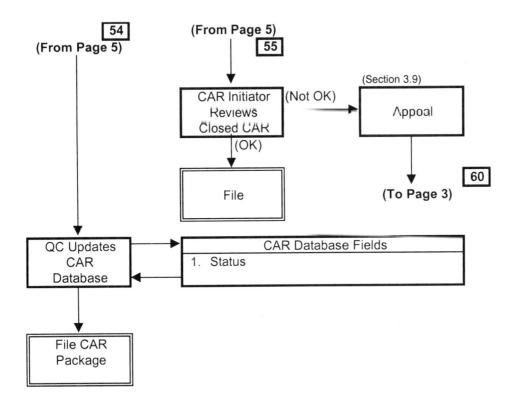

Figure 8-1 Flowchart showing example of corrective action (continued)

The advantage of tracing techniques and process flowcharting is that the decision points stand out visually and the flow of information and decisions are easier to follow than the preceding narrative. Another advantage is that every decision is reduced to a simple *yes* or *no* answer, with one course of action for each.

Trace Forward

As its name implies, a trace forward examines a process from the beginning, moving forward to the end, or moving forward from a specific point in the process to another point further along. The trace forward allows the auditor to get a complete picture of the process from start to finish and to determine the controls of the flow and the consistency of adherence to requirements across interfaces.

When time allows, an auditor will trace a process from start to finish, examining the controls of the flow, especially at critical interfaces, i.e., where a process leaves one department or function and goes to another. These interfaces are the junctures at which most quality system documentation, including checklists, are generated. At the end of a trace, the auditor will have been able to verify process conformity to all standards brought to bear up to that point.

The most typical clauses of the standard that are covered using a trace forward technique are contract review, document control, work control, control of nonconforming products and services, and packing, handling and storage. Although in some situations all clauses will be covered.

The Trace Forward Technique

- ◆ Audits the process from start to finish
- ◆ Determines controls over the process flow
- ◆ Determines consistency of adherence to requirements for the entire process
- ◆ Verifies conformance at interfaces
- ◆ Gives the auditor the whole picture of the process

Trace Backward

The trace backward technique is the reverse of the trace forward. It begins at the end of a process and moves toward the beginning or an earlier point in the process. An advantage of the trace backward is that the auditor begins the trace with a clearer understanding of the expected end result. Using the trace backward technique, the auditor will be able to evaluate the steps in the process in terms of their effective contribution to the desired result. Not only can an auditor determine if processes conform to standards, but the auditor may, in a prescriptive audit, be able to give advice on the need for, or effectiveness of, steps in the process. The trace backward technique is a very powerful tool for verifying the effectiveness of a process and, therefore, a favorite of many auditors.

The Trace Backward Technique

- ◆ Is the reverse of trace forward technique
- ◆ Looks at the process from the end point to the beginning
- ◆ Gives the auditor a clear understanding of the expected results at the start of the trace
- ◆ Allows the auditor to evaluate the process in terms of the effectiveness of reaching the desired result
- ◆ Is a favorite of auditors when verifying effectiveness

The Trace Forward and Trace Backward techniques allow the auditor to determine:

- ◆ Where procedural weakness occurs
- ◆ Where process weakness occurs
- ◆ Where process steps are unnecessary, inefficient, or repetitious
- ◆ At what stage most problems are occurring
- ◆ What is the overall system or process condition

Random Selection Technique

The random selection technique is an alternative that may be used when time and audit personnel are limited. In random selection, the auditor selects various examination points along the process but does not examine between the points. It is during random selection audits that the auditor will have to work hardest to see each part of the process in terms of the whole. In this regard, flowcharts are particularly helpful for showing the complete flow of a process.

An auditor may choose random selection for another reason. After tracing several processes, an auditor may see trends forming. For example, the auditor may decide that the department consistently monitors the quality of raw materials it uses, but is less consistent in its documentation of employee training. The auditor may then choose to focus less on materials quality and focus more precisely on training documentation.

Often, because of logistics, personnel availability, and time limitations, random selection will be the only technique available to the auditor. Usually the random selection type audit will run along the lines of selected clauses of the standard, and at the auditor's discretion, time constraints may be limited to a few of all the standard's clauses. In general, auditors do not favor this technique because the auditor must take more notes so that issues can be verified during the audits and so that audit conclusions can be drawn. Furthermore, it is very difficult for the auditor to get an understanding of the operation and the process flow using this technique. The major reasons the auditor will select this technique over the others is because the auditor needs the flexibility and time savings that the random selection technique provides.

The Random Selection Technique

- ◆ Is an alternative to the tracing techniques
- ◆ Investigates selected points in the process but not the flow
- ◆ Provides a clear focus on a part of the process
- ◆ Is used to further investigate trends uncovered during an audit
- ◆ Is used when time and personnel resources are limited

Functional Responsibility Technique

The functional responsibility technique involves auditing several or all elements within a single department, thereby providing an overall view of the level of implementation and compliance for a particular department. For example, in the manufacturing department, the auditor may ask about the handling of nonconforming products, the corrective action system, quality records, calibration, document control, product identification and traceability, and product handling. In essence, the functional responsibility audit looks at all functions performed by a department or a location.

The functional responsibility technique is often used in small departments, and at remote locations where repeated audit visits would be impractical. When this technique is employed in large offices or departments, it may be combined with random selection, so that an aspect examined in one process is not covered in the next.

The Functional Responsibility Technique

- Is an audit of several processes within a single department
- Provides an overall view of the level of conformance, implementation and maintenance within a department
- Is a very effective technique to use for small offices or remote locations

---

Summary

Whichever technique or combination of techniques is used, the auditor will aim the inquiry at three critical points in any process. These points are the beginning point, or where a group becomes involved with a process; at least one intermediate point; and any point where a process or product is passed from one group to another.

The main idea to remember is that an auditor will have the easiest time proving your department's conformity to system requirements if the auditor understands the goals of each audited process. As has been illustrated here, no combination of audit techniques can guarantee that an auditor will clearly see the big picture. By providing the auditor with flowcharts of the processes, the auditee can greatly assist the auditor and make the audit process run more smoothly.

Chapter 9

COMMUNICATING EFFECTIVELY

A great deal of research goes into an audit. During pre-audit planning, the auditor will have read pertinent standards and procedures that govern a group, and have garnered a good understanding of the mission of the group.

In addition to time spent in research, a significant portion of planning time is spent developing audit questions. Questions are formulated with the goals of efficiency and effectiveness in mind. Efficient questions are short, clear, and concise, allowing the respondent time to answer. Effective questions get the most complete information with the fewest tries.

The Nature of Audit Questions

The auditor reaches two goals with regards to answers to audit questions:

◆ The answers round out the auditor's understanding of the process

◆ The answers lead the auditor to objective evidence of the quality system conformance

Most questions will not be answerable with only a simple *yes* or *no*. The auditor's questions will require thoughtfulness and time to answer.

Every auditor makes the assumption that the people being audited are competent professionals and know much more about the organization's processes than they do. While searching for conformity, the auditor may uncover system weaknesses, and probe those areas more deeply.

But this probe is never personal; the line of questioning should never be construed as pointing the finger of guilt at an individual. The second assumption an auditor makes is that, if a quality system is weak, even the best professionals may be prone to inconsistency in the development, production, and delivery of the organization's products and services.

In a prescriptive audit, where the auditor plays the role of facilitator, questions about a process may go much deeper than they would in a compliance audit. If so, the auditor is attempting to find similarities between the audited department's way of doing a task and another department's way, which may be more efficient or effective.

Prescriptive Audit Questions

Prescriptive audit questions probe to determine quality system weaknesses. They also attempt to find and make parallels to other departments' methods of doing the same thing and share successes with the auditee.

The Nature of Audit Answers

Keep answers brief and to the point. **WARNING!** Do *not* volunteer a long answer when a short answer would satisfy the auditor. There are good reasons for this warning. First, statements an auditee makes will have to be backed up with objective evidence. Second, long answers are habit-forming — the auditee forms the habit of explaining to the point of absurdity, and the auditor forms an expectation of long answers. After hours of long-winded answers, short answers seem suspicious. Finally, the auditee who gives long answers will be prone to drift away from fact and into the realm of opinion, conjecture, and inference, all of which the auditor will attempt to follow-up, but none of which can be substantiated with documentation.

Keep answers objective. Since audit questions are geared toward establishing compliance with system requirements, think in terms of the system and its processes.

The Nature of Audit Answers

◆ Employees must be trained to keep answers:
- Brief
- Objective
- Impersonal
- Professionally courteous

Keeping answers objective and direct may be difficult, even when referring directly to the process instruction. Auditors realize, while auditees might not, that process instructions go through several rewrites, even after being implemented, before they truly reflect a process in all its variety. Hence, what is done in practice may differ from what is written. Answers should not dwell on every exception that has ever occurred or exceptions that may occur in the future. Every process has room for professional judgment.

Keep answers impersonal. Keeping answers impersonal means referring to colleagues by their functional titles rather than names. For example, instead of saying "John proofreads the certificate for typographical errors, and returns it to me," say "The certificate is proofread by the documentation specialist, and returned to the documentation manager." Not only will this rephrasing depersonalize the answer, it will also paint a better picture of flow of control from one functional specialty to another, and demonstrate that the system conformance relies on process control rather than a particular individual.

Keep answers professionally courteous and polite. The auditor and auditee will have occasional differences of opinion about how to interpret a requirement or how to meet it. As a result, an auditee may be surprised by an unexpected turn of the auditor's questions. Keep in mind: the audit is an information-gathering process to verify compliance. At the end, the auditor prepares a detailed report in which the basis for any nonconformity is made clear. The appropriate time for a department to weigh differences of opinion is not during the audit, but in response to the audit findings.

Do not be afraid to ask the auditor to repeat a question, especially if a question seems vague or ambiguous. An attentive auditee who asks the auditor to repeat a question will nearly always get a rephrased, more direct, question. With two versions of a question at hand, the auditee has a better understanding of the auditor's intent. Additionally, the auditee can now choose which question to answer.

Basic Rules to Follow When Answering Auditor's Questions

- Never answer a question you don't understand.
- Never feel compelled to fill silences with long-winded explanations.
- Never invent an answer, especially when a quality record can demonstrate conformance.
- Never answer more than what the auditor is asking.

Trigger Words to Use Carefully

Review	Reasonable
Approve	Negligible
Interpret	May
Exception	Could
Waiver	Should

Dealing with Silence

It may simply be the nature of interviews but, between answers and follow-up questions, ample opportunity exists for silence. Also in the nature of interviews, the interviewee is the most nervous and the most likely to attempt to fill that silence with nervous chatter, and hang himself or herself in the process. Knowing this, experienced auditors go to great lengths to master the art of presenting an expectant face, the kind that looks like a constant question mark, a look that makes the auditee wonder, "Did he (or she) understand my answer?" Don't fall into this trap! Don't hang yourself! Answer the question, and if the auditor really doesn't understand, let the auditor ask another question. Don't volunteer long explanations that are not specifically requested.

Avoiding Confrontations

It is not unusual, when an auditor points out a nonconformance, for an auditee, department manager, or escort to be tempted to defend the practices of the department. Often the auditee will believe the auditor has misinterpreted requirements or is being too stringent.

During the audit is not the time to contest an auditor's views. Any meaningful non-conformances will show up in the auditor's closing report, to which the department will have ample time to devote to a coherent written response. Arguing with an auditor will only sour the trust the department needs to develop between it and the auditor.

Confidentiality

Confidentiality is handled one way during internal audits, and another way during external audits, when the auditor comes from outside the company.

Internal Audit Confidentiality

In an internal audit, both the auditor and the auditee are bound by the same confidentiality agreement, which everyone should review and clearly understand before the audit takes place. In general, auditors are after documentation that demonstrates that a process conforms to requirements. Quality system procedures and process instructions, and quality records, like checklists, are generally considered company confidential. All of these must be laid open to the scrutiny of the internal auditor. The auditor, in turn, has an obligation to keep information confidential within the company.

When an outside customer asks for evidence of quality system conformance and considers internal audit certification sufficient to demonstrate conformance, the written evidence should not report details of company confidential or proprietary information.

An auditee should feel free, at all times, to remind an auditor that specific information provided is confidential in nature and should inform the auditor that copies cannot be made due to the confidential nature of the document.

External Audit Confidentiality

It is common practice for companies, when contracting outside auditors, to agree to specific nondisclosure conditions for the audit. Companies with legal departments or legal advisors on retainer always have counsel review confidentiality clauses of audit contracts for completeness. The agreement should spell out restrictions on an auditor's notes about, or sketches of, a product, explicit naming of suppliers or customers, use of trademarks or proprietary names, and production deadlines or costs. The agreement must be tailored to protect both the auditees company and their customers.

The Issue of Confientiality

◆ Make the confidential nature of information clear to the auditor

◆ Offer alternative samples that do not compromise client confidentiality

◆ Managers should ensure that all employees understand the organization's confidentiality policy

◆ Remember: When applicable, confidentiality is also a requirement which must be complied with

---

In principle, auditors should be given access to quality standards and all evidence of compliance. Access to evidence does not imply the right to direct access to a product or service when direct access is not essential. It is generally considered common practice to restrict access to visual inspection and inform the auditors that copies cannot be made. Keep in mind, an auditor has no basic desire to compromise company security, and will be receptive to suggested alternative ways of verifying compliance other than direct product or service inspection.

Ethics

Finally, and this needs stating, impartiality is the measure of the value of an audit. An audit whose impartiality can be put into question is of no value. Hence, auditors neither expect nor accept bribes, gifts, gratuities, or entertainment. Reasonable refreshments while not expected by an auditor, may be offered when meals will be missed or delayed, or when facilities are remotely located.

Auditors neither expect nor accept bribes, gifts, gratuities, or entertainment.

Chapter 10

FOLLOW-UP AND CORRECTIVE ACTION

The Audit Report

Upon completion of the audit investigation and closing meeting, the auditor will issue a formal audit report identifying each observation and nonconformity uncovered. The comprehensive audit report forms the basis for the corrective action necessary to close the audit and improve the system.

In addition to providing the starting point for resolving nonconformances and observations, the audit report provides the objective evidence that the audit was conducted. As a historical document, the report will be used by the auditees, auditors, and future auditors long after the audit investigation is completed. A sample audit report is presented in Part 3–Reference Material.

Because an audit report highlights only problems, the report is, by nature, negative in tone. This type of report is known as an *exception report*, meaning the auditor cannot report everything that complied with requirements. Such a report only addresses exceptions (nonconformances and observations). The amount of detail that would be required to address everything that was found in conformance with the requirement would make the report the size of a rather large book. Therefore, reports present only the problems.

The report's apparent negative tone must be taken not as an assault, but as genuine, sincere constructive criticism aimed at improving the quality system. Auditors and auditees alike must offer and accept the audit report and findings in the spirit of promoting improvement.

The Audit Report

- ◆ Reports the findings of the audit on an exception basis
  - • Observations
  - • Nonconformances
  - • Summary
- ◆ Is the starting point for corrective action
- ◆ Offered in the spirit of promoting continuous improvement

---

Structure of the Audit Report

The structure of audit reports allows the auditee and senior management to see the level of implementation, maintenance, and effectiveness of the organization's quality system as well as identify areas that require corrective action.

The Structure of the Audit Report

- ◆ Summary
- ◆ Opening and Closing Meeting Registers
- ◆ Audit Scope Statement
- ◆ Office and Organization Description
- ◆ Synopsis of the Audit Approach
- ◆ Summary of Observations
- ◆ Summary of Nonconformances

In general, an audit report includes the following sections:

1. **A summary that provides:**
   a) a report identification number
   b) the names of departments and functions audited
   c) the names of individuals contacted and involved in the audit
   d) the dates of the audit
   e) the names and other identification of the audit team members
   f) a brief concluding statement about the level of implementation, maintenance, and effectiveness of the audited system

2. **Opening and closing meeting registers**

3. **Audit scope stating:**
   a) the extent of the audit
   b) the type of audit conducted

4. **Audit team approach:**
   a) compliance audit
   b) follow-up audit

5. **Summary that identifies observed deficiencies not addressed as nonconformances**

6. **Nonconformance reports or statements**

Audit reports may vary in format and content but all have the same purpose. That is to document what occurred during the audit. The audit report should be duplicated and distributed to each audited department manager. The original should be kept in the organization's files for easy reference.

The audited managers should be tasked to review report sections that address their areas of responsibility and prepare an action plan based on the observations and nonconformances.

Before development of an action plan, the auditor should be contacted to clarify any statements, observations, or nonconformances that are not clearly understood.

Preparing an Action Plan

After reviewing the audit report, each manager should develop a plan for correcting noted observations and nonconformances. The plan should describe the corrective action that has, or will be, taken to prevent the recurrence of the nonconformance or the escalation of an observation to a nonconformance.

An implemented corrective action system is a requirement of almost every quality system, and should be followed in the development of the corrective actions that are described in the action plan. Corrective action forms should be completed for each nonconformance as a method of tracking steps toward resolution. A sample procedure for a corrective action system and a typical Corrective Action Request (CAR) form are presented in Part 3–Reference Material.

A well-formatted action plan contains:

- ◆ A clear reference to the observation or nonconformance being addressed
- ◆ Identification of the root cause of the observation or nonconformance
- ◆ The proposed action to eliminate the root cause
- ◆ The proposed action to implement the corrective action
- ◆ The plan to verify the effectiveness of the corrective action
- ◆ The target date for full implementation and final approval of the corrective action

---

Each manager's action plan should address each nonconformance and observation in terms of the guidelines presented above.

Once managers have completed their action plans, a meeting of the managers or the organization's quality steering committee should be held to discuss each item, with particular attention to common items. The managers or committee should review and approve the corrective action to each nonconformance or observation. One individual should be appointed to consolidate the managers' action plans into a single plan for the entire group. This consolidated action plan is the plan that will be returned to the auditor.

Responding to the Audit Report

In order to close out the audit observations and nonconformances, the auditee must provide the auditor with an action plan. The audit report, or a company's documented procedure for conducting audits, will specify the time frame in which the action plan must be developed and presented to the auditor. It is incumbent upon the audited organization to either respond in a timely manner or to request additional time.

Upon review of the action plan, the auditor will provide comments to the auditee if the plan does not sufficiently address the nonconformances and observations of the audit. If the auditor finds the action plan to be adequate, he or she will specify, in the absence of a written procedure, what is required to close the nonconformances and observations and to close out the audit findings.

Closure of Observations and Nonconformances

An observation or nonconformance may be closed within days, weeks, or months of the audit, depending on when the corrective action is initiated and the degree of effort necessary to preclude its reoccurrence. The auditor may designate a closure deadline. Whenever possible, an observation or nonconformance should be closed as soon as a suitable corrective action has been implemented and verified. Since observations and nonconformances are opportunities for improvement, the auditee should feel a sense of urgency in making the organization more efficient and effective through seizing opportunities to improve.

Closure of Observations and Nonconformances

♦ In general, for the auditor to be able to close a nonconformance or observation, the auditor must be provided with documentation that:

  • States the observation or nonconformance
  • Makes reference to the nonconformance or observation number
  • Demonstrates that the planned actions have been implemented
  • Verifies that the corrective actions are effective in precluding recurrence of the observation or nonconformance
  • The corrective action has been accepted by the organization audited

---

The auditor will review the documentation provided by the organization and will either:

1) Accept the corrective action and close the observation or nonconformance,

2) Request that a follow-up audit be scheduled to verify effectiveness of the corrective action, or

3) Indicate that the corrective action is not adequate to meet the intent of the observation or nonconformance, and provide further clarification.

Final Closure Report

When the auditee has demonstrated to the auditor that all nonconformances and observations have been adequately addressed, resolved, and verified, the closure of the audit will be documented by either the auditors or the organization staff. Managers are urged to share this information with their staff.

Managers should keep audit reports and supporting documentation. In subsequent audits, the auditor will want to examine the results of past audits.

PART 2

INTERNAL AUDITOR TRAINING

*Provides an understanding of the
concepts and techniques needed to
implement an effective internal audit
program and provides the material to
effectively train internal auditors*

Chapter 11

INTRODUCTION

Participation as an internal auditor requires a commitment by participants to help guide their respective company on its journey along the path of quality. In the establishment of this commitment, some commonly asked questions are:

- What is this all about?
- Why did I choose to do this?
- Why was I picked to be an Internal Auditor?
- Where do I start?
- How will this help my organization succeed?

The internal auditor training program is structured to teach and reinforce fundamental knowledge of concepts and techniques that are essential to be a successful internal auditor. The concepts and techniques presented will be what you will encounter and apply to real-life internal auditing situations, which will assist in the development of ability and confidence to conduct internal audits.

Why Perform Internal Audits?

To understand the internal auditor's place in an organization's quality system, we must first know why internal audits are conducted.

Internal audits are an excellent quality system implementation tool. Everyone struggles during the early stages of implementation of a quality system to understand the "new" requirements that quality systems impose. People are unsure of what their role is within the quality system. An internal audit during this stage promotes quality system awareness and provides the organization with the necessary assistance to make the system operate effectively and successfully. These initial internal audits are *prescriptive* in nature; meaning they allow the auditor to act as an implementation facilitator.

As the quality system matures, internal audits help uncover areas where there are opportunities to improve both the quality system and the way the organization does business. At this stage it is critical for an organization to have a structured way of continually measuring how well the quality system is being implemented and maintained. The internal audit process provides this measurement as well as a solid basis for future comparisons.

Very often organizations aspire to become certified to an external standard like the International Organization for Standardization (ISO) 9000 series, which requires internal audits to be performed, without which ISO audit failure is guaranteed.

The ISO 9000 series is not alone in demanding adherence to quality standards. Many businesses worldwide now require that their suppliers demonstrate conformance to an internal quality management system as a condition of doing business with them.

Internal audits are also an excellent verification tool. Through them, management can discern the answers to questions including:

♦ Is there conformance to quality system standards or other standards?

♦ Is the quality system understood?

♦ Has the quality system been fully implemented?

♦ Is the quality system being maintained at all levels?

♦ Does the quality system meet the needs of customers, the organization, and other changing standards?

♦ Does the quality system provide effective control to assure conformance?

Internal audits are also a management tool. Using questions such as stated above and data collection techniques to verify conformance to quality system requirements, *auditors verify that an organization does what it claims to do.* Hence, internal audits are based upon the organization's own quality system requirements which may be in accordance with the ISO 9000 series standards.

If internal audits are based solely on an internal quality system, how does a company meet the ISO 9000 series requirements? Organizations who intend to pursue certification to one of the ISO 9000 series quality standards design their own systems around those requirements. Therefore, an internal or external audit of the organization's system verifies that their system complies with relevant clauses of the ISO 9000 standard. Similarly, when a company or organization does business that requires compliance to a specific standard, they comply by incorporating the relevant aspects of the standard into internal requirements.

Internal audits are an independent and objective means to verify an organization's activities. Audits do not attempt to verify that technical specification or production methods are necessarily good. For example, an audit will not answer the question, "Should we use these industry standards or others?" The audit can only test adherence to an expressed standard.

Types of Audits

There are as many types of internal audits as there are reasons for performing them. The type of internal audit performed will be based on the objectives of the internal audit. The objective may be to measure the quality system as a whole or just a particular element of the system. The auditor may wish to examine a particular procedure or conformance to a set of requirements or a specific process, or may want to focus on a particular product or service the company provides, or a particular function or operating group. Finally, the auditor may conduct a follow-up internal audit to confirm corrective actions prescribed by a previous audit.

Types of Audits

- System audit
- Procedure audit
- Process audit
- Product/Service audit
- Function audit
- Follow-up audit

The objectives of the audit will lead the auditor in selecting a particular type of audit to conduct. Each audit type takes a look at verifying conformance to the requirements from a slightly different perspective, but all have the same goal. Below is a brief description of the various audit types an internal auditor may select.

System Audit - A system audit would be used to determine whether or not the quality system or one or more of its elements has been developed, implemented, and maintained in conformance with the established internal requirements or with the requirements of an external standard.

Procedure Audit - A procedure audit verifies that the procedures meet the requirements of the system and that work associated with the procedure conforms to the requirements of the procedure. Technical expertise in the procedure being audited may be necessary.

Process Audit - A process audit verifies that the process conforms to the established requirements and generally verifies the process from initial process activity to the final process activity. The scope of a process audit is generally much narrower than that of a system audit.

Product Audit - A product audit focuses on a particular product or service provided by the company and determines whether or not the requirements for the product or service have been complied with from the initial process through the final process. As with the procedure audit, technical expertise may be necessary to conduct a product audit.

Function or Department Audit – A function audit is utilized to determine whether a function or department conforms to the requirements for the processes for which they are responsible. This type of audit is the most common for small companies or offices within a company.

Follow-up Audits – A follow-up audit is performed to verify that the corrective actions resulting from nonconformances issued during a previous audit have been effective in preventing reoccurrence of the nonconformance. These follow-up audits are generally short and directed only at verifying resolution of corrective actions.

Prescriptive and Compliance Audit Methods

All of these internal audit types may be conducted using either the prescriptive audit method or the compliance audit method. The auditor must decide which method best fits the needs and objectives of the organization and the goals of the audit. The maturity of the quality system is often the driving factor behind choosing a method.

The Prescriptive Audit

The prescriptive audit method used during the implementation phase of the quality system allows auditors to increase company awareness of the quality system requirements, to assist in system implementation, and to measure how well the quality system is performing. During a prescriptive audit, the auditor plays two distinct roles, that of fact-finder, or detective, and that of teacher, or facilitator. While fact-finding is an important aspect of all audits, in prescriptive audits, teaching is also important. Why?

In the early stages of a quality system, the auditee may have a poor understanding or low awareness of the quality system requirements. The auditee may, for instance, have well-written procedures, which show how responsibility is shared and what standards are expected, but then fail to write process or work instructions, which show the step-by-step controls in a task.

In the prescriptive teaching role, the auditor is in a good position to provide calm, objective guidance, to clear up misunderstandings and to inspire motivation. To be most effective, the auditor must be intimately familiar with the organization's goals, the quality system requirements, and the management objectives. Done properly, a prescriptive audit should be an enjoyable challenge and a rewarding experience, because the auditor plays an active role in ensuring the success of the quality system.

The Compliance Audit

In all types of audits, the auditor is on a fact-finding mission to discover whether the organization complies with the requirements it has established for providing products and services. As a quality system matures, prescriptive audits decrease in frequency, and are replaced by internal compliance audits. In contrast to a prescriptive audit, in which the auditor is expected to assist in resolving problems uncovered during fact-finding, an internal compliance audit is best defined as a strictly fact-finding mission to verify conformance to established requirements.

In a fully mature quality system, organizations use compliance audit methods exclusively. The compliance method is closely aligned to the methodology employed in external certification audits. The chart of Figure P2-1-1 illustrates the typical progression, over time, from prescriptive to compliance auditing methods.

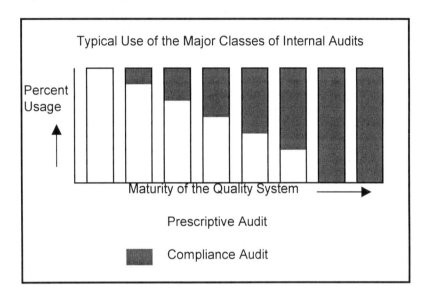

Figure P2-1-1 Typical progression from prescriptive to compliance auditing methods

Preventive Focus

Prevention is one of the central objectives of internal audits. Quality system audits allow organizations to identify opportunities to improve processes and the quality system before problems with either become serious enough to negatively affect the company and the product or services that it provides. The internal audit is one of the most powerful tools an organization has to uncover these improvement opportunities.

The internal audit is an effective way to help the organization improve. That's its purpose, and the auditor must communicate that purpose clearly. The message the internal auditor must convey is "We are here to help you improve and keep your processes under control." By clearly focussing on prevention, the auditor fosters an atmosphere of continuous improvement and creative change in the organization.

A major question all organizations facing an external certification audit ask themselves is "What will the external auditors find that does not comply with our requirements?" With an internal audit program based on prevention, the organization will find and solve its own problems before they're found by an external auditor. It is always easier for an organization to take the necessary corrective action when it does not have to deal with the additional pressures of attaining or maintaining certification.

Although the internal benefits of the preventive nature of internal audits are important, the most significant benefit is gained when an organization finds problems before they become evident to the organization's customers. An organization is bound to lose hard-won, hard-held customers unless it can identify potential problems and opportunities for improvement before customers and competitors do.

Internal audits, which are focused on prevention, are carried out because of their positive impact on the organization's culture, customers and, ultimately, "company profitability."

Chapter 12

AUDIT PREPARATION

This chapter provides information that will assist you in:

- ◆ Developing audit plans and schedules
- ◆ Determining audit scope requirements
- ◆ Assembling an effective audit team
- ◆ Conducting pre-audit team meetings
- ◆ Carrying out reviews of quality system documentation
- ◆ Notifying auditees

When an organization commits itself to an internal audit program, it sets about developing procedures to outline the requirements for preparing and executing audits and reporting results. A sample section of a quality system manual, internal auditing procedure and related process instructions are shown in Part 3 – Reference Material.

Internal Audit Preparation

Auditors face many issues during internal audit preparation. The time the auditor spends in preparation is critical to the audit's success.

To ensure that everything will run smoothly on the day of the audit, an auditor should consider the following facets:

- ◆ Plan
- ◆ Scope
- ◆ Team selection and make-up
- ◆ Quality system documentation review
- ◆ Scheduling
- ◆ Notification
- ◆ Pre-audit team meeting
- ◆ Checklist

All aspects of internal audit preparation, except audit checklist development, are covered in this chapter. Audit checklists will be covered in Chapter 13.

The Audit Plan

The best way to guarantee the success of an internal audit is to prepare an adequate audit plan. The first step in this process is to identify the departments or functions to audit and the key individuals to contact. Once the departments, functions, and key individuals have been identified the auditor can proceed with audit plan development.

The Audit Scope

- ◆ Study/review function/department to be audited
- ◆ Identify processes and quality system elements
- ◆ Identify internal requirements
- ◆ Identify external requirements

The scope of the audit is determined by identifying all of the processes and the quality system elements in which the audited function of department is involved. Preparing the audit plan and defining the scope of the audit early will help the auditor determine the type of audit to conduct.

Regardless of the type of audit to be carried out, all internal and external requirements that apply to the audit must be identified. The internal requirements may include the organization's quality system manual and quality system procedures. External requirements may include industry or government standards such as ISO, Mil-Std., or EPA. External requirements will indicate the extent to which a function must be audited, which will help you, as the auditor, finalize the audit plan and decide how many days to devote to the audit. The audit plan, scope, and length of the audit all depend on the amount of material the auditor has to cover. Figure P2-2-1 shows an example of an audit plan and scope.

Audit Schedule – April 6 - 7

Date	Time	Function	Auditee
Monday, April 6	8:00 – 8:30	Opening Meeting	All
	8:30 – 10:00	Human Resources	J. Smith
	10:00 – 11:00	Marketing	B. Martin
	11:00 – 12:00	Sales	A. Thomas
	1:00 – 2:00	Engineering	P. Gonzalez
	2:00 – 3:30	Shipping	K. Phillips
	3:30 – 4:30	Quality	T. Jones
Tuesday, April 7	8:00 – 9:00	Management	F. Morgan
	9:00 – 10:30	Finance	S. Malone
	10:30 – 12:00	Design	A. Gomez
	1:00 – 2:00	Environmental	G. Johnson
	2:00 – 3:30	Business Develop.	P. Todd
	4:00 – 4:30	Closing Meeting	All

Figure P2-2-1 Example of an audit plan and scope

Audit Team Selection and Make-up

Internal audits may be conducted by individuals or by audit teams. The number of auditors required for an audit depends on the nature and scope of the audit and the amount of time available to perform the audit.

When the nature of the audit suggests that an audit team would be most effective, the team members must be carefully considered. The prospective team members' qualifications, experience, and availability are all determining factors in the selection process.

For internal audits, the team should include auditors with a background in auditing, members with knowledge of the quality system requirements, and members who are specifically familiar with the technical focus of the department or function being audited. If the scope of the audit involves specific analyses of statistical elements or mathematical calculations, then an auditor competent in these areas should be included in the audit team as well. All members of the team should, however, have knowledge of the auditing process.

An audit team is composed of a team leader (lead auditor), experience auditors, and possibly auditors-in-training. Each audit team member has a specific role.

Lead Auditor Responsibilities

- ◆ Organize and control the audit
- ◆ Identify nonconformances
- ◆ Evaluate corrective action
- ◆ Select team

The lead auditor is responsible for organizing and controlling the audit, identifying nonconformances, and evaluating the corrective action taken as a result of previous audits. The lead auditor is responsible for selecting the team members who should be selected from a pool of auditors certified to the organization's internal auditing certification requirements or certified by an external certifying body such as ABS. Part 3 – Reference Material contains an example of the process carried out for the internal certification of auditors.

Audit Team Unity

Since the audit team members have to work closely with one another during the audit, the compatibility of the auditors is very important. Each auditor must respect the other's role and work supportively during the audit. The audit team required attributes include unity, compatibility, objectivity, and integrity.

Audit Team Integrity

Auditors must be independent of the department or function being audited to ensure objectivity. The objectivity of the auditor is critical to a successful audit. Depending on the size of the organization, the team leader may face a significant challenge simply trying to locate an auditor with the needed technical expertise and auditing background who is independent of the function being audited. Despite this challenge, the auditor's objectivity and independence from the function must be assured prior to assignment to the audit team. Some additional but very important considerations in audit team member selection are the personal and professional integrity of the auditors, and their ability to follow the ethical standards of the organization.

Whether the audit is conducted by an audit team or by an individual auditor, it is important to adhere to the preparation steps and associated thought processes to ensure a successful internal audit.

Quality System Documentation Review

Audit Documentation Review

- ◆ Quality system manual, procedures and industry standards
- ◆ Previous internal/external audit reports
- ◆ Previous nonconformances and corrective actions
- ◆ Past/current corrective action requests

With the audit's scope defined and the audit team selected, the next step is for the lead auditor to obtain all documentation related to the system or process being audited. These documents may include, but are not necessarily limited to, the organization's quality system manual, procedures, and industry standards. The lead auditor must, together with the other team members, review and become familiar with the requirements of these documents. Previous internal and external audit reports must also be reviewed to identify specific subjects or activities to be covered during the audit. The auditor should also review the results of corrective actions to previous nonconformances, as well as past and current corrective action requests, to determine whether or not to verify their follow-up during the audit.

As these documents are reviewed, audit team members should take notes on the requirements of the particular departments, functions, processes, and system components, and consider each element for inclusion in the audit checklist.

Scheduling of the Audit and the Notification of the Auditees

Audits should be scheduled and the auditee notified in a timely manner. When developing the schedule for the internal audit, consider the following factors:

- ◆ Preparation time requirements
- ◆ Scope and complexity
- ◆ Number required to conduct the audit
- ◆ Availability of auditors/auditees
- ◆ Conflicting activities

During the scheduling phase there will, of course, be informal communication between the audit team and the auditees, during which a tentative audit date and schedule may be set. Once a firm date has been established, however, leaders of the audited department, function, or process should be officially notified in writing. Auditees should be notified at least one month before the audit date to allow the department sufficient time to prepare for the audit and to notify individuals who must attend.

Pre-Audit Team Meeting

Prior to the audit, the audit team should meet informally to discuss the audit objectives, clarify team members' roles, and plan the strategy and methods to be used during the audit. The audit team should also review the requirements to which the auditee will be measured for conformance, the specific auditor assignments within the departments and functions to be audited, and any team member concerns. Arrangements for meeting on the day of the audit should also be made at the pre-audit team meeting.

Chapter 13

DEVELOPMENT
OF AUDIT CHECKLISTS

The information in this chapter will help you:

> ◆ Determine the purpose of audit checklists
> ◆ Develop audit checklists
> ◆ Determine the checklist type and format

The Purpose of Audit Checklists

The audit checklist can be used as a road map, an efficiency tool for the audit, and a detailed record of the audit.

As a Road Map

Trying to conduct an audit without well-developed checklists is like going on a long journey without a map. The end result is always wrong turns, frustration, frayed nerves and, worst of all, a waste of everyone's time and effort. Auditors use checklists to set goals and to pin down details of key elements they would otherwise be apt to miss. Therefore, it is crucial that an auditor prepare audit checklists before each audit. An audit checklist is developed in such a manner that each objective of the audit is known and can be met, so that all standards to which compliance must be measured are covered.

As a Tool of Efficiency

An auditor is responsible for covering a lot of often unfamiliar ground quickly, so efficiency is essential. The time the auditor spends developing checklists pays dividends during the actual audit. By being prepared and knowing what to look for in advance, the auditor will project an image of competence, efficiency, believability, and professionalism. In short, the auditor will gain the auditees' trust.

As an Objective Record

A checklist provides the objective record used by auditors on which to base their conclusions and substantiate their findings. During the course of an audit an auditor talks with many employees. A checklist documents which elements conform and which do not. Relying solely on memory, an auditor would have a difficult time composing a complete and accurate report. Checklists put all of the audit details right at hand during the report writing stage.

As a Standard

Often a single process is performed the same way or to the same standard everywhere within a company. For the sake of efficiency, and to ensure consistent conformity to a common standard, a single checklist is developed with which all departments may be measured. A good example would be a checklist for auditing an organization's company-wide use of its *corrective action system*, department by department. The consistency of a single checklist provides a particularly important tool to monitor the continuous improvement of the quality system. As standards change, the single list measures by one yardstick how the change is being implemented company wide.

Steps in Checklist Development

Step 1: Determine Objectives and Scope of the Audit – Before specific processes can be identified for auditing, the auditor must clearly understand the scope and objectives of the audit. To determine the scope and objectives of the audit an auditor needs to answer some fundamental questions.

- ◆ Is this a compliance or prescriptive audit?
- ◆ Is the entire system or specific elements to be audited?
- ◆ Is one function or all functions to be audited?
- ◆ Will the audit follow the processes from start to finish?
- ◆ Will the audit follow the product/service from initial stages through delivery, installation and servicing?

Step 2: Identify Processes and Related Requirements – Step two in developing an internal audit checklist is to identify the processes to be covered in the audit. Once the objective and scope of an audit are known, the auditor can determine which system elements will provide items for the checklist. In most companies the system elements and their related requirements are defined in the following documents, from which the auditor must:

◆ Review quality system manual, procedures and process/work instructions
◆ Identify processes covered and standards
◆ Review any contractual requirements
◆ Develop checklist questions based on requirements review

The auditor reviews these documents, identifying processes to be covered and standards for each, so that questions can be developed for the checklist.

Additionally, the auditor investigates whether contractual obligations exist that would require verification of conformance to standards that are not part of the company's quality system — particularly important when a company's product or service is audited. The auditor would request that the auditee provide copies of contractual requirements early in the audit planning stage so that these additional requirements can be included in the audit checklist.

During the requirement review stage of building a checklist the auditor gains an understanding of the flow and interrelationship of the processes that will be audited. The development of the checklist in this manner allows the auditor to get the big picture of how the organization should function.

Step 3: Develop the Questions – A critical step in the development of a useful checklist is painstaking formulation of the questions. The auditor needs to answer the following questions when developing the internal audit checklist:

◆ What documents and evidence should be reviewed?
◆ What points to check?
◆ Which department is responsible for meeting requirements?
◆ Which individual can verify conformance?
◆ How can question be phrased to ensure clarity and brevity?

By answering these questions the auditor will be able to create questions that are meaningful, arranged in a logical order and which ensure complete coverage of the system.

Step 4: Determine the Sequence – Once the audit objectives and requirements are understood and meaningful questions have been developed, review the questions and arrange them meaningfully, so that the audit flows.

The checklist needs to follow a logical sequence that actually described the requirements in the order they are being met. A lack of logical order to a checklist makes it very difficult to ensure complete coverage of the system.

An example of well thought-out questions for verifying the conformance in a typical purchasing and procurement department would be:

- ◆ Are procedures prepared and maintained for the distribution of purchasing documents, change orders, and the purchase orders established?
- ◆ Have all employees who are involved in the process trained in the use of the procedures, and are the requirements understood by the employees?
- ◆ Is there a current list of approved suppliers available?
- ◆ Do the purchase orders sent to suppliers include all of the requirements requested by the manufacturing department?
- ◆ Are the purchase orders and change orders reviewed and approved prior to release to the suppliers?

The above example covers all the relevant requirements for purchasing and procurement and follows a logical sequence from development of the requirements to the release to the suppliers?

- ◆ User friendly
- ◆ Space for questions, criteria and comments
- ◆ Include identifying information
- ◆ Align with selected audit technique
- ◆ Keep it simple

Step 5: Audit Checklist Format – The internal auditor has the freedom to choose a format that best facilitates the performance of the audit. In all cases the chosen format should be user friendly and have sufficient space for the questions, criteria and comments. Equally important, the checklist must have the basic audit identifying information, including date, location, subject, audit identification number, who the auditor is, the employees contacted, and other information that will assist the organization's understanding of the relationship between the checklist and the actual internal audit.

As discussed in the previous section, it is critical that checklists follow the intended flow of the internal audit. This means that whichever method is chosen for an audit, the checklist must be aligned with the techniques of that method. Three audit techniques, random selection, trace forward, and trace backwards, are discussed in Chapter 8. The importance of an organized and structured checklist will be more evident after these audit techniques are explored.

Checklists do not have to be very complicated. Simple formats are often the best because auditors are comfortable using them. It is also important to format a checklist so the auditor can readily check off items as they are verified, thereby leaving very few, if any, blank questions. If the checklist is designed in this manner, it will help the auditor ensure a complete audit and a complete, orderly audit report.

Questions Versus "Memory Joggers"

Auditors often wonder if it would be acceptable to just jot down memory-jogging notes about requirements, rather than developing full-blown questions. The use of short memory-joggers is a good technique for someone who is very familiar with the requirements to which the audit is being conducted. The memory jogger statement checklist method should generally be reserved for auditors who have experience asking questions and probing so that the requirements stated in the memory joggers are fully investigated for conformance.

The biggest benefit of a memory jogger type of checklist is that it can be developed in a modular format. Each auditor within an organization can select the modules to be audited, thereby customizing a checklist for the audit. If a complete memory jogger type checklist is developed it can be used effectively as an "off-the-shelf" checklist by other experienced auditors within the organization.

A drawback to the "off-the-shelf" memory jogger checklist is that the auditor is generally not as prepared as an auditor who builds a full-question checklist by reviewing the documentation before the audit and different auditors may interpret the checklist differently.

Pros -   Modular format
             Can be used as "off-the-shelf" checklist by other auditors

Cons -   Auditor generally not as prepared
             Different auditors may interpret differently

Examples – Various style checklists have been included in Part 3 – Reference Material, to provide a flavor of what a checklist should look like. One of these formats may be adopted by your organization or you may develop you own.

Chapter 14

AUDIT TECHNIQUES

This chapter will help you:

- ◆ Understand the importance of communication
- ◆ Ask effective audit questions
- ◆ Manage problematic auditees
- ◆ Understand the different investigative techniques
- ◆ Manage an audit team
- ◆ Separate fact from inference and opinion

It is vital for an auditor to have a solid grasp of the choices of techniques that may be used to conduct an audit. The success of an internal audit rests on the auditor's ability to communicate effectively, to ask questions that aid in verifying conformance, to listen attentively to the auditee, to observe the auditee and the surroundings, to be able to employ varied investigating methods during an audit, and to write effectively.

Effective Communication

From time to time in our daily associations with friends, customers, and coworkers communication breaks down and misunderstandings happen no matter how hard we try.

Effective communication is always a challenge and it is no less so in audit situations. Different people hear different messages from the same words spoken by the same speaker, such as when a manager states that a task is *challenging and rigorous*. One worker may hear this as "hard and picky" while another may hear it as "ambitious and demanding."

> **Helpful Communication Techniques**
>
> ◆ Use auditees' language
> ◆ Speak at auditees' educational level
> ◆ Solicit feedback
> ◆ Rephrase/repeat to ensure comprehension

As a matter of startling fact, researchers say that only one-third of a speaker's message is actually received and understood by a listener, one-third is either garbled or not understood, and the remaining third of the message is not heard at all! To make matters worse, different individuals will actually hear and understand different parts of the intended message.

The Spoken Word May Not Be Heard

The first major hurdle to communicating effectively is ensuring that the listener hears what is being said. Various impediments stand in the way of a listener's ability to hear what is said. These impediments include physical barriers such as distance from the speaker, noise, and disinterest on the listener's part, poor concentration, and tiredness. The auditor must try to foster an environment that eliminates as many of these elements as possible.

It is good practice, whenever possible, to conduct the audit in a quiet setting that takes the auditee out of the path of day-to-day interruptions. Providing a quiet setting may not always be possible, particularly in a manufacturing facility. It remains the auditor's responsibility to have the message heard by the auditee.

What Is Heard May Not Be Understood

Misunderstanding is another barrier to effective communication. The listener may hear the spoken words clearly, but utterly fail to understand what the speaker is saying. Misunderstanding in spoken communication is a complex phenomenon that may have roots in several causes. One of the more common causes of misunderstandings include an individual's knowledge or lack of knowledge of the subject under discussion. The speaker's vocabulary may include jargon (words with which only the experts in a subject are familiar), or the speaker and listener may have differing degrees of familiarity with the language. The listener may, for example, be uncomfortable with the speaker's regional accent. Differences in educational level between the speaker and listener can also play a role in communication effectiveness.

Helpful Communication Techniques

The auditor has techniques at his or her disposal that go a long way toward ensuring that what is heard by the listener is what is understood. An auditor who makes an effort to use the auditees' language and vocabulary and to speak at the auditees' educational level vastly improves the chances that the message will be understood. Another effective technique is to solicit feedback from the auditee. Ask the auditee if he or she understands. As an auditor, do not be afraid to rephrase your message if you think the auditee did not understand it the first time. Remember, as an internal auditor, your goal is to help and misunderstandings certainly do not serve this goal.

What Is Heard and Understood May Not Be Accepted

Denial is another barrier to effective communication. For several reasons a listener may not be prepared to accept what he or she is being told.

There are two potential barriers to effective communication: listener denial and personal prejudices.

Personal prejudices are the most difficult obstacle for an auditor to overcome and can only be handled on a case-by-case basis. An auditor will find that he or she learns to deal with prejudice through experience.

As if these barriers alone were not enough to confuse issues, the auditor faces even further difficulty. Words and preconceptions are only part of the overall communication process. Face-to-face communication is also greatly affected by *nonverbal aspects* of the message transmission process. In fact, nonverbal signals are at times so strong they completely obliterate the spoken message. An auditor must be aware of the way nonverbal elements of communication affect how a message is heard, understood, and accepted.

Body Language

There are two distinct ways that people communicate nonverbally. The first is through body language, which includes posture, movements, facial expressions, gestures, and the physical distance we keep between ourselves and others (interpersonal distance). Through the use of body language distinctly different messages can be sent. For example, good posture sends a message of confidence, while bad posture can send a message of timidity and indecisiveness.

Tone and Style

The second way we communicate nonverbally is through vocal tone, inflection, and delivery. An auditor's vocal tone, inflection, and delivery may be misinterpreted by an auditee as patronizing and judgmental. The auditor must exercise care at all times to ensure that the words, body language, vocal tone, inflection and delivery all convey the same message and meaning. Later in this chapter, we will learn how an auditor spots and interprets the nonverbal signals sent by the auditee.

It is important that auditors convey the intended message and confirm that the auditee understands the message. The auditor must also listen intently to the auditee and strive to hear and understand the auditees' messages. These are difficult tasks that the auditor will master with experience.

Questioning Techniques

It would appear that an auditor's job is to go around asking a lot of questions. While this is true in part, the more significant and most difficult part of an auditor's job is to be a good listener and gather factual information upon which to base a conformance assessment. A good auditor will apply the 80/20 rule with 20 percent of the time spent asking questions and 80 percent of the time spent listening to the answers. The quality of the auditees' responses depends in great part on the skill with which the question is asked. Therefore, an auditor must have a sound mastery of questioning techniques if he or she expects to collect all of the needed information.

Effective Questioning

- Structure line of questioning – THINK
- Put auditee at ease
- Vocal characteristics and delivery
- Interpersonal distance
- Comfort zone
- Unresponsive auditee
- Evasive answers
- Confrontation

Questioning is the most essential ingredient to an effective audit. The ability of an auditor to properly structure the line of questioning is the key to facilitating the needed communication between the auditor and auditee. Therefore, it is important to ensure that the thought process behind the questions is properly organized into subject categories and information flow patterns before conducting the audit.

Think Before Asking Questions

Even though the auditor with a prepared audit checklist has an excellent aid to structuring a line of questioning, it is critical to make sure "the brain is in gear before operating the mouth." A checklist is only an aid, not a script written in stone. Communication is give and take, and an auditor will often need to improvise. An auditor will need to be coherent going into the audit *and mentally* prepared to ask questions.

Putting the Auditee at Ease When Questioning

An auditor must understand that many people, especially auditees, find the audits stressful. If the auditor is relaxed, there is a greater likelihood that the stress level of the auditees will be less. Being relaxed means the auditor should be "at ease" while still being professional. As the auditee responds positively, relaxes and becomes more comfortable with the situation, the auditor will notice that the auditee is more willing and better able to provide the appropriate responses. This is particularly important since the only way an auditor can successfully determine compliance with the requirements is through the openness and cooperation of the auditee.

Vocal Characteristics and Delivery Relative to Questioning

The nonverbal communication skills of the auditor play a major role in influencing the attitude and stress level of the auditee. Of particular importance are the vocal and nonverbal characteristics of the auditor, how the questions are delivered, and the interpersonal distance.

The way an auditor says something matters greatly as to how it is received by the auditee and how the auditee responds. The most unpleasant things spoken in a gentle tone and manner can be well received by the auditee, while pleasant things said in a harsh tone can be taken negatively. How the auditee perceives what the auditor says depends on such factors as the pace of speech, vocal pitch, volume, and tone of voice. These factors can mean that certain styles of speaking and accents are more readily accepted than others and that some styles should be shunned.

The following example illustrates how vocal characteristics and delivery affect perception.

In some parts of New York and the Northeast people are accustomed to speaking quickly and abruptly. They are comfortable being answered the same way. Such rapid-fire delivery probably would not be accepted in the South because there a quick delivery is considered aggressive. On the flip side of the coin, Southerners often complain of not being taken seriously in the North because of their accents and manner of speaking. In the North this slower, more deliberate way of speaking is thought to indicate a slowness in the thought process as well.

Interpersonal Distance

The distance we stand from one another when talking is known as *interpersonal distance*. This real but often ignored or underrated factor in one-on-one communication, should be given thoughtful consideration by an auditor who will be involved in extensive one-on-one communication over the course of an audit.

The Comfort Zone

At some point along the route to becoming civilized, each of us establishes something called *personal space*—the space and physical barriers we need around ourselves to feel comfortable when we communicate with others. The borders of this space are often called the *comfort zone*. Depending on setting, our audience (the closer the relationship, the smaller the comfort zone), our personal experiences, and the culture in which we grew up, our comfort zone may range from a few inches to several feet between speaker and listener.

On an unconscious level, we have all been aware of this concept all our lives. This is illustrated by the term 'close friend' that means just that—a friend who is allowed inside the protective border that strangers may not cross. *Close* is a relative term. What is close to one person may be too far to another. People raised in cities generally have a much closer comfort zone than people who grow up in the country. When speaking with an acquaintance, the city dweller's comfort zone may be twelve to fifteen inches, whereas a country speaker's comfort zone may be four feet or greater. Comfort zones also vary from country to country and from culture to culture. In Britain and Germany, for example, speaker and listener generally stand farther apart when talking than Americans would, while in Japan, Mexico, and the Arab countries, speaker and listener stand closer together.

It is critical that an auditor develops sensitivity to an auditee's comfort zone. If the auditor invades that zone, the auditee will be uncomfortable, may feel intimidated, and may not be able to concentrate on what is being said. By being inside an auditee's comfort zone, an auditor builds, rather than breaks down, barriers to communication. Fortunately, someone whose comfort zone has been invaded will have a natural tendency to move away from the invader. An auditor must learn to sense when this is happening and back off. Even after the auditor backs off, however, the communication lines will remain strained for a time until the perceived threat subsides. Therefore, it is better to take, in advance, whatever precautions are necessary to avoid violating an auditee's comfort zone.

Manage Problematic Auditees

During the course of most audits, situations will arise that have the potential to become awkward if not dealt with quickly and effectively. In questioning a nonresponsive auditee, or one who gives evasive answers, or in trying to deal with a confrontational auditee, the auditor may feel growing frustration at his or her ability to perform the task at hand. To succeed, an auditor must be proactive, meaning the auditor must learn to spot a potential barrier and to take steps to prevent the barrier from becoming real.

Dealing With Unresponsive Auditees

An auditee may, for varied reasons, be deliberately or unintentionally unresponsive. Some, for example, are easily intimidated by even the gentlest of auditors, because they view the auditor, who sits in judgment, as an authority figure who holds the balance of power over the auditee's career. Another auditee with career concerns may want to hide a problem the auditee *knows* is not his or her fault, out of fear the auditor will not view the problem in the same light. The auditee feels that by saying nothing they will not be blamed. Whether the cause is deliberate or unintentional, the auditor must learn to sense such situations and plan to interview others involved in the function or element. If the auditor *does* interview others, the interviews must be handled tactfully. Quite often the most probable alternative choices are subordinates of the unresponsive auditee.

Dealing With Evasive Answers

An auditor may encounter auditees who give very evasive answers. Evasion generally has two root causes. First, some auditees try to impress the auditor with their "vast" knowledge of the subject. The second and more common cause is an auditee trying to talk his/her way around a question because of a *lack* of knowledge about the subject.

An auditor should learn to spot this situation quickly. A good auditor, while not accepting what the auditee is saying, must remain calm, polite, and tactful. To solve the problem the auditor can restate a question more specifically to allow the auditee to provide either a short, specific answer or to simply say they do not know the answer. When rephrasing a question it helps, psychologically, to give the auditee the benefit of the doubt—to assume the auditee simply did not understand the original question.

If restating a question fails and an auditee continues to give evasive answers, the auditor may need to ask the escort, management representative, or department manager to assist in clarifying the question to the auditee or to provide another employee who can answer the question. As a last resort, the manager may be directly asked to answer the question.

Dealing With Confrontation

It is not unusual, when an auditor points out a nonconformance, for an auditee, department manager, or even an escort to try, sometimes heatedly, to defend the practices of the department. Often the auditee will claim the auditor has misinterpreted requirements or is being too picky. In the worst case, the competence of the auditor will be called, vocally, into question.

It is important that an auditor remain calm, polite, self-controlled, and in control of the audit. Flexibility is often the way around such confrontations. It is possible to remain in control and be flexible at the same time. Try to take a different tack with the auditee. A good way to do this is to say to the auditee "Let's review the situation so that we all understand whether a problem really exists." By remaining poised and professional an auditor should be able to diffuse the situation.

Because an auditee may feel the auditor is raising a nonconformance against them personally, the auditee may also believe company management will blame them personally for the problem. Therefore, it is important that an auditor make clear to the auditee, with occasional reminders, the purpose of the audit, which is not to assign personal blame, but to verify whether the controls that are in place are adequate to ensure conformance to system requirements. Additionally, the auditor should emphasize that an audit looks at systems, processes, and products—not individuals. If a confrontation occurs that eludes a quick resolution, the auditor must, early on, accept that the auditee is not going to back down and simply move forward with the audit. The auditor will have ample opportunity to raise the nonconformance at the closing meeting and in the audit report. Don't get trapped in a debate in which there will be one winner and one loser. Audits are not debates. No purpose is served by pursuing a confrontation if sufficient evidence has already been gathered.

Learning to Listen

Although learning to ask questions is an important skill, listening skills are just as, if not more, important. Remember during pre-audit preparation, the auditor has the luxury of time in choosing questions and encoding them. The auditee, in contrast, has no such luxury. Auditee's answers may be incomplete and disorderly, sentences fragmented. Considering all of the normal barriers to communication and the additional ones posed by the audit environment, an auditor must have a good mastery of listening skills. Listening has two part: hearing and understanding what is being said and observing and understanding what is being nonverbally communicated.

To be a good listener an auditor must stay focused and learn to deal with the noise created by the unfamiliar surrounding's activities.

During an audit an auditor must absorb, process, and evaluate a great deal of information quickly. Additionally, an auditor must be able to distinguish between the important and the unimportant, as well as separate fact from inference and opinion. These responsibilities demand that an auditor be attentive. Sooner or later, however, every auditor will find him or herself in the position of sheepishly asking an auditee to repeat an answer because the auditor was not paying attention. The only way to minimize these unfortunate circumstances is by knowing why they happen.

Inattentiveness

An auditor must be wary of falling into traps that lead to inattentiveness. Two pitfalls that commonly plague auditors are *preoccupation* with planning the next question and *deliberation* of the previous question's answer. Either of these can cause the auditor to miss important information, both verbal and nonverbal. If an auditor becomes aware of falling into either of these traps, the auditor may choose to increase the time between questions, thereby taking the time to listen, understand, and process answers at a pace that best facilitates getting the information required.

Inattentiveness has another cause: preconception — believing you know what an auditee's response will be before you have heard it. Recall that internal audit teams usually have members with extensive knowledge of the processes of the department being audited although such knowledge may not be up to date. As a result an auditor may so strongly expect a familiar response to a question that the auditee's actual answer goes unheard. Preconception is a double-edged sword. An auditee may hear the start of a question and, believing he or she knows how the question will end, will stop listening and begin preparing a response.

Daydreaming

Studies have shown that the average person understands speech delivery at a rate of up to 600 words per minute, while the average person's delivery falls somewhere between 100 and 140 words per minutes. This difference between the rate of delivery and understanding causes another common barrier to effective listening that is daydreaming. Daydreaming happens when a listener has time to think about something that is more interesting or entertaining while the speaker slowly plods along.

To become a good listener, which is a must for an auditor, the auditor needs to practice concentration and attentiveness. An excellent exercise for building concentration and comprehension skills is for the auditor to occasionally repeat back, using other words, the auditee's answer, asking for confirmation that the message was correctly understood.

Listening Behavior

There are various types of behavior that affect an individual's ability to listen effectively. Listeners can be categorized as:

> ◆ Defensive listeners
> ◆ Insensitive listeners
> ◆ Communication hogs
> ◆ Pseudo listeners

Defensive Listeners

Defensive listeners listen for something more than what is said. Like readers who read between the lines in a written report, the defensive listeners pick apart what is spoken to find a nonexistent hidden meaning and then defend themselves against the message they thought they heard. Additionally, a defensive listener may go on the offensive, fighting to preempt the message they expect to hear next.

When an auditor feels an auditee is a defensive listener, the auditor should either keep the questions short and simple, so the auditee won't have time to read anything into the question, or pose questions in direct, explicit terms whose intent cannot be misinterpreted.

When, as a listener, an auditor feels himself or herself going on the defensive, use the same technique you would use to guard against inattentiveness. Restate the auditee's answer in other words and ask for conformation that the meaning was correctly understood. If there *is* a difference between what the auditee meant and what the auditor understood, the auditee will certainly clarify the statement given.

•

Insensitive Listeners

Insensitive listeners, not attuned to nonverbal signals in communication, rarely receive the whole message. Insensitive listeners often appear disinterested or distracted. The auditor must strenuously avoid the trap of insensitivity, which will quickly sour the trust between the auditor and an auditee and bring cooperation to a halt. Auditors will encounter occasional auditees who are insensitive listeners. While there are no absolute solutions to the problem, an auditor may increase the chances of effective communication by making and waiting for, if necessary, eye contact. One note of caution: Some people are, in general, uncomfortable with prolonged eye contact. Watch for nonverbal signs of this. Another technique that helps is to address an auditee by name (first or last, as appropriate), as in "John, (eye contact) what do you do when you receive an invoice from...?". One note of caution on this, too. If you use an auditee's name, get it right.

Obviously, if you do these things and still find you're getting nowhere, tactfully and professionally pursue your line of questioning with someone who *will* listen intently.

Communication Hogs

Communication hogs talk constantly, often saying very little or saying the same thing several ways. The communication hog only allows the others to speak while they are catching a breath before they fire their next volley. The communication hog appears to have little or no awareness of what others are saying or whether others are listening. This is an especially bad trait in an auditor, since an auditor's goal, really, is to *get the auditee to speak* about their systems and processes.

Since communication hogs often take a great deal of time to say very little, this type of behavior is also understandable in an auditee. Because an auditor's time is limited, action must be taken. There are two options an auditor can pursue. The first option is to thank the individual for answering the questions and moving on to someone else in the department. The second opinion is to pose only *yes or no* questions. While these might seem, on first thought, inadequate, they are clearly preferable to complete communication paralysis.

Pseudo Listeners

Have you ever heard the expression "The lights are on, but no one's home?". Pseudo listeners nod and respond at the appropriate times, but it quickly becomes evident that their thoughts are elsewhere. An individual who starts out really listening may adopt pseudo listener behavior if he or she perceives that what is being said is either boring or common knowledge. The attentive auditor should spot this shift and attempt to stimulate the auditee's interest by rephrasing or changing the way questions are asked. On the other side of the coin, an auditor should not become a pseudo listener. An auditor who only pretends to listen will ultimately impress the auditee as uncaring. One of the principal reasons an auditor must not fall into this trap is that auditees say very important things that should not be missed, even when what the auditee is saying merely restates what is common knowledge, as is often the case in an audit. The only thing an auditor can do here is focus on the auditee and pay attention. It may not be easy, but it *is* necessary.

Learning to Observe

To be an observant auditor and to be able to draw logical conclusions, attention to detail and an ability to think analytically are needed qualities. However, conclusions drawn on the basis of observations should not be reported until substantiated by objective evidence. The conclusions drawn from observation generally should be used as signposts to problems.

For an auditor to properly use observation as a tool he or she must understand what observation entails. Observation is:

> ◆ Receiving messages
> ◆ Interpreting communications
> ◆ Having an inquisitive outlook
> ◆ Visually noticing the obvious as well as the not so obvious

Body posture and movement, facial expression, and gestures are common forms of nonverbal communication that can be easily misunderstood, misinterpreted, or simply just missed. It takes practice and experience to observe nonverbal signals in the correct context so that conclusions can be drawn. Luckily, we observe people every day of our lives and all develop some ability to spot and interpret these signals. The auditor must hone these skills to get the most out of the audit investigation.

The following examples illustrate how body language affects an auditor's impression of the auditee:

> ◆ How would you compare an auditee who sits sprawled across a chair, to one who sits upright during an audit?
> ◆ What would you think about a fellow who claims to enjoy his work but answers your questions with a hanging head and slumped shoulders. Would you believe this claim?
> ◆ Facial expression conveys a great deal about what a person is thinking and feeling. What message would you get from the woman who told you, while wrinkling her brow and frowning, that everything in her department was running smoothly? What if she were smiling?

Auditors must also be aware that auditees are observing them. With this in mind, an auditor must be careful not to send the wrong signals with facial expressions or body language.

Gestures

Gestures can also be expressive and even dangerous. The same gesture can have very different meanings in different settings. A gesture that is harmless in one place can be considered extremely rude in another. The auditor must always be mindful of this and gestures that may alienate the auditee.

It is critical that an auditor, while observing the auditee, realizes that audit questions often uncover broader problems or weaknesses than one would at first suspect. The first hints of these problems may be observed as nonverbal signs. Close observation of nonverbal clues may assist the auditor to arrive at a finding that is closer to the root cause of a system deficiency.

Effective Writing

The final deliverables an auditor leaves with an auditee are the written *Audit Report* and *Nonconformance Statement*. These documents are the basis upon which an auditor's credibility is either validated or completely destroyed. When writing the report and nonconformance statement, it is of paramount importance that an auditor be factual, concise, complete and, above all, correct.

It is essential that an auditor take detailed notes during an audit. Audit notes, used in conjunction with the audit checklist, are an auditor's best resource for developing the audit report and nonconformance statement. Accurate note taking is also the basis for reconstructing the audit events. Notes supplement the auditor's recollections and discussions with the audit team and serve as a point of reference when following up a lead. Notes may, in fact, be an auditor's only record of an audit, especially since the use of audio and video tape recorders is almost universally frowned upon.

The most important single thing an auditor must remember when writing is to never sacrifice correctness, completeness, or clarity for the sake of brevity.

When auditing on a team basis one team member can be designated as the official note taker, supplemented by the lead auditor's own notes.

Investigative Techniques

Since the goal of an audit is to collect objective evidence through investigation that measures conformance to a system, process, or product to the stated requirements, sample documents and even products may be reviewed.

Four Commonly Used Audit Techniques

- Sampling
- Tracing
- Random selection
- Functional responsibility

Sampling

An auditor has only a limited time to verify conformance during the actual audit investigation. This time constraint makes it impossible for an auditor to examine every document that might attest to the control of the system. Therefore, an auditor must take conclusions based on representative samples. Generally, a few samples are adequate to ascertain the status and level of conformance of a product or process. The auditor should randomly select a handful of samples from the top, middle, and bottom (or front, middle, and back) of a container or file cabinet to get a representative cross section. Auditees often expect the auditor to take samples from the top of the pile or front of the cabinet only, and may actually try to hide problems in the middle or bottom (middle or back) of the container or file cabinet.

The auditor should not fall into the trap of complacency, just picking samples from the top or allowing the auditee to select the samples. The best place to take process samples is at the interfaces of the process or product (i.e., when the product goes from one department to another).

It is critical to record pertinent data about the sample so that samples can be related to data that substantiates the nonconformance as fact. The auditor should record the following information about a sample:

- Location from which the sample was taken (stage of the process, department, machine, etc.)
- Part/document name and number
- Lot number or other significant identifiers
- Time and date the sample was taken
- Variable or attribute results
- Any observations that may affect the process or product
- Whether or not a nonconformance should be raised

Plans for sampling should be prepared before the audit, although plans may not always be followed because the auditor does not know exactly what will be discovered during the audit. For internal audits, a standardized sampling plan may be practical.

Tracing

Tracing, a highly effective audit investigation technique, is an orderly examination, by sampling at each stage, of the flow of a process or system for effectiveness, conformance, and traceability. Flowcharts often help in the tracing technique. In order to reveal the complete flow of a process, tracing may require an audit of several departments. There are two distinct methods of applying tracing techniques: trace *forward* and *trace backward*.

Trace Forward

As its name implies, a trace forward examines a process from the beginning, moving forward to the end, or moving forward from a specific point in the process to another point further along. This method is quite effective since the auditor gets a complete picture from start to finish and can determine the controls of the flow and consistency of adherence to requirements across interfaces.

Trace Backward

The trace backward technique is the reverse of the trace forward, moving from the end of a process toward the start or to an earlier point in the process. The advantage of the trace backward is that the auditor begins with a better understanding of the end objective of the process. All steps in a process can then be evaluated in terms of their effective contribution to the desired result. The trace backward technique is a very powerful tool for verifying the effectiveness of a process.

Both trace forward and trace backward techniques allow the auditor to determine:

- Where procedural weaknesses occur
- Where process weaknesses occur
- Whether steps are unnecessary, ineffective, or needlessly redundant
- At what stage in the process problems occur
- The overall condition of the system, process, or product

Random Selection

The random selection technique is an alternative to the tracing techniques that may be employed when time and personnel are limited. In the random selection technique the auditor selects various points along the process flow chain but does not trace the process between the selected points. Often times because of logistics, ongoing projects, availability of personnel, and time limitations this is the only technique available to the auditor. The disadvantages to this technique are that the auditor must take more notes so that issues can be verified during other audits and conclusions can be drawn and it is difficult for the auditor to get an understanding of the operation and process flow. The advantages to the random selection technique are that it allows the auditor more flexibility and saves time.

Functional Responsibility

The functional responsibility technique is another method used by auditors in verifying conformance to the requirements. This technique is usually reserved for the more experienced auditor because it is more encompassing. It involves auditing several, often many, elements within a single department. For example, when auditing the manufacturing and assembly department, questions may be asked about handling of nonconforming products, the corrective action system, quality records, calibration, document control, product identification and traceability, and handling as they relate to the department. The advantages to this technique are that it covers several elements in one location, provides an overall view of the level of conformance for a particular department, and only disrupts the activities of the function being audited.

Effective Use of Audit Teams

Often times the scope of an internal audit is large enough that more than one qualified auditor is needed to successfully complete the audit in the allotted time. In these situations the audit team approach is employed. This section addresses how an audit team is used effectively to conduct a large scale audit.

Lead Auditors and Seconds

Certain factors must be taken into consideration when deciding which team member will audit which element. These include the auditor's system knowledge, the auditors work experience, and technical expertise. The assigned audit roles may also help to determine what elements are assigned. For example, it might be more suitable for the lead auditor to conduct the audit of senior management when verifying management commitment or management review. The key role in addition to conducting the actual audit investigation is the pre-audit preparation and post-audit reporting. The lead auditor also acts as the final authority during the audit and facilitates the successful completion of the audit. The auditors will either audit assigned system elements, processes, or products independently or assist the lead auditor during the investigation.

Team Auditing vs. Individual Auditing

The actual audit investigation and interview may be conducted either individually by a single auditor or by a team of auditors. Some of the advantages of work as a team are that one member can ask the questions and focus on the response given while the other member is free to observe the activity, and document the responses, mannerisms, and body language of personnel in the surrounding area. Sometimes, an auditee may ask someone on his or her staff to retrieve data or records not available in the immediate area. Working as a team allows the second auditor to accompany the staff member while the first auditor remains with the auditee. Whether working as a team or individually, this tactic can be useful in assuring that the data or records presented remain objective and do not skew the sample or results of the investigation. Other advantages are that a team can achieve a more comprehensive audit and it is the preferred method of training an inexperienced auditor. The major disadvantage is that the auditee may be more intimidated by a team of auditors than by a single person.

Follow-up

Because audit styles differ from auditor to auditor, the amount of time expended during an investigation also differs. If one auditor completes his investigation before the other auditor does, he may follow up on behalf of another auditor where weaknesses or potential problems may exist. Additionally, when two auditors are assisting each other in one department, if a situation arises which indicates that a problem may be present in another department or rooted in another department, one of the auditors can continue the audit of the department while the other verifies the problem in the other department. This saves time and improves the continuity of the audit process.

Sounding Board

During auditor team meetings, auditors should discuss areas of concern with the rest of the team so as to gain additional opinions on interpretations and implementation by the auditee. This can either confirm or validate a previous position or provide sufficient clarification to resolve the uncertainty.

Facts, Inferences, and Opinions

Considering that an auditor's responsibility is to obtain objective evidence and report factually on the findings of the audit, it is critical that the auditor has the ability to separate the facts from inferences and opinions. The auditor must be able to develop this skill when listening, speaking, and writing.

When Listening

The auditor must be able to ascertain the context in which something is said. If the auditee states, "This is how we should do this procedure...", he or she may be stating a fact. On the other hand, he may be inferring that the procedure is not actually performed as documented or then again, he or she may be giving a personal opinion on how the procedure should be performed. If unsure of the context of the response the auditor should ask again.

As important as it is to listen to what is said it is equally important to listen to how it is said. Listen for tonality and inflection. Listen, don't just hear. Accept a confession but be prepared to verify a claim.

When Speaking

An auditor must be clear and concise when speaking and questioning. The auditor should not ask leading questions that can infer one thing when it really means something else. The auditor is much like a detective on a fact-finding mission. All the auditor is looking for is "the truth, the whole truth, and nothing but the truth." The auditor is not on a fault finding crusade asking questions like a prosecuting attorney trying to discredit a witness. An audit is an investigation to verify conformance to the requirements, not an inquisition or a witch hunt. An auditor should be prepared to hear different interpretations of how the requirements are satisfied. It is important for the auditor not to let his or her established paradigms and background lead the questions. Stick to asking questions that verify conformance. It must be explicitly clear as to when an auditor should give opinions and when the auditor should refrain from doing so.

When Writing

As with speaking, an auditor's writings must be clear and concise. Nonconformances must be clearly stated and written. The auditor must not infer that a nonconformance exists if there is no objective evidence or data to substantiate the claim. When stating facts, avoid the use of "should" or "could." When giving opinions and recommendations, they should be listed separately and titled as such. It should also be noted that the auditor's opinions and recommendations are just that, opinions and recommendations. The opinions and recommendations of the auditor may not be the only way to achieve the desired goal, which is to conform to the requirements.

TYPE	ADVANTAGES	DISADVANTAGES
TRACE FORWARD	• SHOWS LOGICAL SEQUENCE • EASY FOR TRAINING • DETECTS FRONT-END WEAKNESS QUICKLY • AIDS IN PRE-PLANNING OF AUDITEE PERSONNEL	• NOT VERY FLEXIBLE • NOT PRACTICAL FOR PARTIAL AUDITS • FLOW BROKEN IF AUDITEE PERSONNEL NOT AVAILABLE
TRACE BACKWARD	• STARTING POINT OPTIONAL • EASY FOR TRAINING • RESULTS OF WORK SEEN PRIOR TO AUDIT • AIDS IN PRE-PLANNING OF AUDITEE PERSONNEL	• NOT VERY FLEXIBLE • FRONT-END WEAKNESS NOT AUDITED UNTIL THE END • FLOW BROKEN IF AUDITEE NOT AVAILABLE
RANDOM SELECTION	• VERY FLEXIBLE • GOOD FOR PARTIAL AUDITS • GIVES BROAD PICTURE QUICKLY • MINIMIZES DISRUPTION • AVAILABILITY OF AUDITEE PERSONNEL CAN BE JUGGLED	• NOT GOOD FOR TRAINING AUDITORS • REQUIRES EXPERIENCED AUDITORS • REQUIRES GOOD NOTES • MORE DIFFICULT TO UNDERSTAND FLOW
FUNCTIONAL RESPONSIBILITY	• SHOW OVER-ALL UNDERSTANDING OF THE QUALITY SYSTEM BY DEPARTMENT • COVER SEVERAL ELEMENTS IN ONE LOCATION • VERIFY SYSTEM IMPLEMENTATION	• NOT GOOD FOR TRAINING AUDITORS • REQUIRES EXPERIENCED AUDITORS • REQUIRES GOOD NOTES

Figure P2 – 4-1 Audit Inspection Techniques

Chapter 15

CONDUCTING THE AUDIT

This chapter will help you:

- ◆ Conduct both the opening and closing meeting
- ◆ Verify compliance to requirements and identify non-conformances through questioning, data collection, and analysis
- ◆ Understand the difference between conformance, observations and nonconformances
- ◆ Conduct and participate in audit team meetings and auditee debriefings
- ◆ Write meaningful nonconformance statements

The Opening Meeting

The opening meeting is an important part of the audit and can have a significant impact on the success of the audit. This is the audit team's opportunity to make a good first impression and to establish the tone of the audit. A typical agenda for the opening meeting will contain the following:

- ◆ Introduction
- ◆ Audit method
- ◆ Purpose of the audit
- ◆ Scope of the audit
- ◆ Timetable and agenda
- ◆ Defining of nonconformance and observations
- ◆ Auditor roles
- ◆ Logistics and resources
- ◆ Questions and answers

For small offices or departments the opening meeting can be handled on a more informal basis.

Verifying Compliance and Identifying Findings

After completing the necessary groundwork to prepare for the audit and after presenting the schedule, scope, audit methodology, and purpose of the audit to the auditee, the internal auditor is ready to venture out into the workplace to begin the audit investigation. In so doing, the auditor is armed with the auditing techniques learned in Chapter 4, the checklist, and a note pad.

The auditor's primary responsibility during the on-site conformance verification phase of the internal audit is to uncover as many facts as possible that demonstrate that the auditee is actually doing what they say they do to assure that their activities meet their requirements. An auditor will apply various investigative techniques to verify compliance. These techniques include:

- ◆ Observing, listening, and questioning
- ◆ Collecting pertinent records
- ◆ Probing
- ◆ Questioning
- ◆ Sampling records
- ◆ Listening
- ◆ Tracing
- ◆ Following up on leads and claims

To carry out such an audit the auditor will be talking with many employees, including the organization's upper management. Remember the auditor is in control of the situation and must ask the necessary questions and probe to verify conformance. If the auditor has prepared properly, the auditees will be more nervous than the auditor will. The auditor should always remain calm, cool, and collected, making the job easier and helping to put the auditee at ease. Reducing the nervousness of the auditee will make it easier to get the evidence needed to assess the system.

The compilation of the objective evidence that either verifies that the auditee's process or system element conforms to the requirements or uncovers the existence of an observation or nonconformance is a difficult task that requires concentration and tenacity.

Important Definitions

Before the auditor begins the actual audit, it is important that the auditor has a clear understanding of the definition of the terms *fact, objective evidence, observation, and nonconformance or nonconformity*.

Fact – Something told by the auditee to an auditor that is seen by the auditor, or something documented which is taken to be true.

Objective Evidence – A documented statement of fact or other record pertaining to an item or activity based on information, observations, measurements, records, or tests that can be verified.

Observation – A weakness detected in an element in the quality system which, if not corrected, may result in a degradation of the product or service quality and possibly could become a nonconformance.

Nonconformance or Nonconformity – A departure of a quality characteristic from its intended level or state that occurs with a severity sufficient to cause an associated product or service not to meet a specified requirement.

Helpful Do's and Don'ts

- Look around and get a general impression
- Feel free to talk to employees other than the department manager and supervisors
- Do not read checklist as a script
- Base conclusions on factual and objective evidence
- Let the auditee do the talking
- Be a good listener
- Take comprehensive notes
- Avoid asking questions that have a "yes" or "no" answer

When the auditor arrives at the first location to be audited, he or she should look around to get a general impression of how the area is maintained and setup. At this time the escort will introduce the auditor to the manager of the department and the audit begins. In almost all instances the auditor will also have to question other members of the department to get all the answers needed in order to draw conclusions. The auditor should feel free to talk to not only the department manager but also the supervisors and any other employee. The success of the audit hinges around auditing where the activities take place and talking to the people who carry them out.

The auditor does not have to ask the questions as written on the checklist. The checklist questions are really questions to the auditor. An auditor will have a difficult time observing the auditee and developing a flow if he or she is trying to read a script. This will also give the auditee the impression that the auditor is rigid, does not know what to look for, and is not very well prepared to conduct the audit.

For the auditor to be satisfied that the auditee has provided the necessary factual evidence to demonstrate that the system either complies with or does not comply with one of the checklist questions, a series of questions may be necessary and a number of documents reviewed. It is important to remember that an auditor must base the auditor's conclusions must be based on factual and objective evidence.

Often times an auditor will find it difficult to obtain the factual and objective information from an auditee. This is especially true when the auditee knows he or she has something to hide or when the auditee feels that he or she may be open to criticism if nonconformities are found. The communication techniques presented in Chapter 4 should be employed to eliminate or abate these situations.

The usefulness of the answers solicited from the auditee greatly depends on how the questions are asked and is the key to combating potential problems. Phrase the questions so that they do not lead the auditee to the correct answers and so that the auditee's answers are descriptive. The more the auditee talks the more information the auditor will be able to obtain and the greater the chances are that the facts will come to the surface. By letting the auditee do the talking the auditor will have the opportunity to probe and uncover the needed facts. To get the most out of the answers given, the auditor needs to be a good listener and note the key points and leads expressed by the auditee. Therefore, it is important that the auditor avoid asking questions that can be answered with either "yes" or "no." The best way to phrase questions which will help you verify conformance is to ask, "Why does..., How do you..., Where do you..., When do you..., Who is responsible for ..., and What do you...."

Even though the objective of the audit is to verify conformance the auditor will discover nonconformances.

Ask questions to determine if:

- ◆ The auditee complies with specified requirements
- ◆ The control is effective
- ◆ There is documentation
- ◆ The process or product flow is traceable from initial to final activity

---

Although the auditee may satisfactorily answer the question the auditor asks, the auditor needs to ask for and review samples of documentation that verifies that what the auditee claims is being done. This can best be referred to as the "Show me..."or "Could I see"... request.

When an auditor discovers or suspects a nonconformance:

- ◆ Request and review samples of documentation–"Show me"
- ◆ Remember that a nonconformance is usually an opportunity to improve a work process.

Do not let the auditee go off alone and bring the sample documents he or she wants to show as objective evidence. The auditor or another member of the audit team should go with the auditee and select the samples to review. Take sufficient notes regarding the documents reviewed as they could be used as objective evidence if a nonconformity is uncovered. This will assist the auditor in justifying a nonconformance and helps the auditor develop a meaningful nonconformance statement. When the auditor discovers a nonconformance the auditee should be told that a nonconformance was found which will be raised in the audit report.

There will be other times when the auditor suspects that a nonconformance has been uncovered but is not completely sure. Let the auditee know that a problem may exist. This gives the auditee the opportunity to clarify the situation. If the auditee demonstrates that a nonconformance does not really exist, the auditor should move on in the audit. If after giving the auditee the opportunity to clarify things, the auditor feels that the factual and objective evidence exists to substantiate the suspected nonconformity, state that a nonconformance has been uncovered and that it will be raised in the audit report.

There will be times when the auditee will disagree with the nonconformance; give the auditee the opportunity to discuss it. An auditor should never get into arguments with the auditee over a nonconformance. Since the auditor is in control of the situation, back off, and take a different tack. It is also important to avoid getting into a discussion of the best solution to a problem unless a prescriptive audit is being carried out.

The key to effectively managing the audit and avoiding arguments at the closing meeting is to have the department manager agree to the existence of the opportunity for improvement. If the auditor has based the nonconformance on the facts, the department manager will most likely agree with the auditor without arguing.

There may be times during the questioning that an auditee will admit that they do not do something that is required by a procedure or the system. Take this admission of a nonconformance as fact and inform the auditee that a problem has been uncovered which will result in a nonconformance statement being raised. Getting agreement from the auditee in this situation is easy since the auditee has admitted a recognized shortcoming.

Often times while auditing a particular department the auditor will identify potential problems in another department due to the close interaction and dependency of the various departments in the organization. It is not necessary to interrupt the conformance verification of the department when a potential weakness in another department has been uncovered. Just make a note that the issue needs to be followed-up, either when the auditor is through with the department being audited or during the scheduled audit of the other department. Additionally, a nonconformance uncovered in the department being audited may actually have its root cause in another department. This should also be noted so that the nonconformance can be verified and raised against the correct department or departments.

Upon completing the questioning in the department, thank the department manager and other departmental employees for their time and cooperation. Then, with the escort, move on to the next area to be audited and start the process all over again or ask to be taken to a department where a lead that has been uncovered during the audit can be followed up.

Audit Team Meetings

Audit team meetings allow the audit team to discuss the progress of the audit in a private setting. The communication among the audit team is important for a successful audit. Audit team discussions can provide the following:

- Opportunity to reach consensus
- Consistency in audit flow
- Better understanding of requirements
- Forum to discuss and categorize findings
- Discuss any disagreements in private
- Review complex issues and documents

These meetings should be scheduled daily at a specified time, either in the morning before starting the audit investigation or in the afternoon upon completion of the day's audit investigation, depending on the audit timetable. The meetings should be brief, approximately thirty minutes long. The entire audit team (lead, second, trainee, etc.) must attend for this to be an effective meeting. During these meetings the auditors can begin drafting the observations and nonconformances, and keep abreast of the developments of the audit. Depending on the audit progress, the timetable may be revised for effectiveness an deficiency. The cooperation and communication of the audit team is the key to a successful audit.

Auditee Debriefing Meetings

- Important element for a successful audit
- Strengthen rapport
- Can be scheduled daily
- Opportunity to discuss any areas of concern or problems
- Changes to timetable

The debriefing meetings with the auditees are also very important for a successful audit, especially for audits that cover many system elements and departments. Debriefing meetings strengthen the rapport the auditors have with the auditees. These meetings can be scheduled daily and can act as a daily closeout by reviewing the findings from the activities covered that day. It is also an opportunity to discuss any areas of concern or problems from the completed activities. Changes to the timetable may also be discussed during these meetings.

Writing Meaningful Nonconformance Statements

- Documentation
- Objective evidence
- Specified requirement
- Corrective action

In addition to being able to identify a nonconformance based on the factual evidence obtained during the audit investigation phase, the internal auditor has to be able to convey the nonconformance to the auditee in an effective manner. Verbal communication when the nonconformity is uncovered must be followed up with a written nonconformance. These nonconformances must be completed and given to the auditee before leaving the facility.

A nonconformance statement is the documentation of the objective evidence which identifies a departure of a quality characteristic from its intended level or state that occurs with a severity sufficient to cause an associated product or service not to meet a specified requirement and is of sufficient detail to allow initiation of corrective action. The key words in this definition are: documentation, objective evidence, specified requirement, and enable corrective action.

<u>Documentation</u> - The written nonconformance statement is a vital piece of documentation. A written statement will ensure that the nonconformance can be understood and the evidence collected by the auditor can be used in corrective action.

<u>Objective Evidence</u> - The nonconformance must be factual and based on something the auditor sees, something the auditor is told by the auditee as being company policy, company requirement, or something or part of a requirement. The auditee needs to know how and where something was uncovered so that the evidence can be used to determine the root cause and resultant corrective actions.

<u>Specified Requirement</u> - The auditor is required to identify to the auditee which requirement is not in compliance so that the corrective actions can be developed.

<u>Corrective Action</u> - The auditee is required to take whatever action is deemed necessary to prevent the problem from occurring again. The nonconformance statement is the starting point for corrective action and therefore must be understood by the auditee.

Nonconformance statements must be written so that all parties involved in the audit understand the nature of the problem as it relates to the requirements.

Attributes of a Well-Written Nonconformance Statement

- Limited to a few sentences
- Factual and correct
- Understood by both auditor and auditee
- Includes statement of requirement associated with nonconformance
- State objective evidence
- State where nonconformance was identified
- State how nonconformance was uncovered
- State nature of nonconformance

What a Noncomformance Statement Should Not Be

- An auditor's opinion of how things should be done
- Long complex dissertations on the reasons a particular problem occurred
- Suggestions for corrective action
- A conclusion based on the auditor's impression of the auditee

Since the nonconformance statement is the basis by which an organization improves the way they do business, the internal auditor must master writing nonconformance statements in order to help the organization take corrective action. If an internal auditor keeps in mind the goal of helping the organization improve while writing the nonconformance statements, it will be easier to make them meaningful.

Closing Meetings

> ◆ Ensure auditee fully understands all observations and nonconformances
> ◆ Schedule immediately after conclusion of audit and final team meeting
> ◆ All participants should attend

The closing meeting is the last opportunity for the auditors and auditees to get together to review the activities of the previous days and results of the audit. The main objective of the meeting is to ensure that the auditee fully understands all observations and nonconformances identified during the audit. This meeting should be scheduled immediately following the conclusion of the audit and the final audit team meeting. The auditor should recommend to management that all individuals who participated in the audit should attend the closing meeting. A typical agenda for the closing meeting will include the following:

> ◆ Review purpose and scope of audit
> ◆ Rules of meeting
> ◆ Summary of observations and nonconformances
> ◆ Review auditees' responsibilities
> ◆ Audit report
> ◆ Questions and answers

Review the Purpose and Scope of the Audit

This will be a brief summary of the extent of the audit outlining the processes covered and why the audit was being conducted. During this time the attendance register should be signed by all present.

Rules of the Meeting

The lead auditor will define the format of the meeting so that it is effective and efficient. The auditor may choose to review all the findings and then have discussion, or have discussion after reviewing each finding. These are the rules that need to be clarified at the beginning of the meeting or problems will develop during the meeting.

Summary of Observations and Nonconformances

The lead auditor should review each finding with the auditee as described when reviewing the rules. The lead auditor also has the responsibility of controlling the meeting by keeping any discussions brief and to the point. As the findings are reviewed it is necessary to have the auditee acknowledge receipt of each nonconformance. The auditee may not agree with the nonconformance; however, acknowledgment of the nonconformance is necessary and serves as evidence that the nonconformance was issued to the auditee. This can be accomplished by simply signing the nonconformance.

Review Auditees' Responsibilities

The responsibilities of the auditee after the audit team leaves must be reviewed. The need for a follow-up action plan, which is generally required by the auditees' procedures; the response date to the auditor, and the requirement of follow-up plans, are usually established within thirty days of receipt of the audit report.

The Audit Report

The auditor should inform the auditee of the target issue date of the audit report. Additionally, the auditor should provide the auditee with a summary evaluation of the audit and advise if a follow-up audit is recommended. The summary evaluation should provide the auditee insight into the tone and message that will be conveyed in the audit report.

Questions and Answers

The auditee must be given the opportunity to ask questions or clarify any issues since that will be the last time all the audit participants will be together.

The auditors should close the meeting by expressing their thanks for the assistance provided during the audit.

Chapter 16

AUDIT REPORT
AND FOLLOW-UP

In this chapter, you will learn to:

> ◆ Prepare an effective and comprehensive audit report
> ◆ Take post-audit actions
> ◆ Provide auditee with documentation needed to close out the audit

The Audit Report

The audit report is the document that provides the objective evidence that the audit was conducted and will be used and referenced by the auditees, the auditor, and future auditors after the audit visit is complete. This document should reflect the preparation and the results of the audit in a format that is thoroughly comprehensible.

It is recommended that the first sheet of the audit report be a summary sheet. This summary should provide important information such as:

> ◆ The report identification number.
> ◆ The location or address of the department or office audited.
> ◆ The individuals contacted and involved in the audit.
> ◆ The date(s) the audit was conducted.
> ◆ The audit team members
> ◆ A summary of the findings

The body of the report will follow the summary sheet contain the following information:

> ◆ Scope of the audit
> ◆ Office organization
> ◆ Audit approach
> ◆ Audit team conclusions
> ◆ Summary of observations
> ◆ Nonconformance statements

Scope

Describe the extent of the audit noting the processes or system elements that were audited. State the type of audit that was conducted and identify the particular departments or office audited.

Office/Organization

Describe the organization of the departments or office and its reporting lines within the organization. Identify the functions performed by the employees in the departments or office audited and their interface with the rest of the organization.

Approach

Describe whether the audit was a prescriptive, compliance, or follow-up audit. Provide any relevant information to support the approach.

Audit Team Conclusions

Describe the level of conformance to the audited requirements. Note in general terms the areas of conformance for the department or office. State whether a follow-up audit is or is not recommended and why. A follow-up audit will be determined by the number and severity of nonconformances identified during the audit and the level of implementation of corrective action.

Summary of Observations

List the areas where deficiencies were identified but were not addressed by nonconformance statements. For each item, it is important to provide the basis by which the auditor determined that the deficiency was present. Source documents should be cited whenever possible. The auditor should include all observations considered applicable and constructive.

Nonconformances Statements

The nonconformance statements are included in this section of the audit report. For each nonconformance statement, it is important to reference the requirement and source document.

Question, Answer and Follow-up

- ◆ Question and answer session to ensure all aspects of report are understood
- ◆ Auditee is responsible to prepare an action plan within a specified time
- ◆ Follow-up and corrective action implementation
- ◆ Follow-up audit as may be required
- ◆ Nonconformance statements
- ◆ Final closure report

When the completed audit report is distributed to the auditee, the auditor should be available to answer any question that may arise from the report. Since the report is not a comprehensive account of the audit, clarification may sometimes be needed from the auditor.

Upon receipt of the audit report, the auditee is required to prepare an action plan to address the observations and nonconformances noted in the audit report. The action plan should describe the corrective action that has been taken or will be taken to prevent the recurrence of the nonconformance or observation. This action plan should be submitted to the auditor or the manager of the function responsible for assuring implementation of corrective actions within a specified period from the date the audit report was received. The auditor or manager of the function responsible for assuring implementation of corrective actions will then verify that the corrective action complies with the nonconformance statements as written and effectively resolves the problem of conformance to the original requirement. Based on the review of the action plan, the action plan is accepted or returned to the auditee with comments. The auditee will then revise the corrective action and resubmit the action plan.

The responsibility for follow-up and corrective action implementation usually does not rest with the auditor. Management may assign the director of quality, a quality steering committee, or a quality improvement team the responsibility for assuring corrective actions have been taken to prevent the nonconformance from reoccurring.

A follow-up audit may be required to verify that the proposed corrective action has been implemented and that the possibility of the nonconformance recurring has been eliminated.

Nonconformance Report Closure

A nonconformance statement describes the requirement and the circumstances relating to why the requirement is not satisfied. The nonconformance can be closed after the auditee, or any other group responsible, initiates corrective action. The auditor can accept closure to the nonconformance after review of documentation supporting the initiation of corrective action, or after a follow-up audit to verify the nonconformance has been effectively resolved.

The NCR can be closed within days, weeks, or months of the audit depending on when the corrective action is initiated and the extent of the effort necessary to preclude its reoccurrence. All nonconformances should, if possible, be closed prior to the next regular audit of that department or office.

Final Closure Report

The auditor has the responsibility to acknowledge receipt of the auditee's action plan. When the auditor or function responsible for the implementation of corrective action is satisfied that the proposed corrective action will adequately resolve the nonconformance, the auditee must be notified. This response can be formal or informal and should be documented in a letter or in a report when the nonconformance has been cleared.

Chapter 17

AUDITOR QUALITIES AND CERTIFICATION

This chapter provides:

- ◆ A list of personal and professional attributes and management skills needed to be an auditor
- ◆ Example of internal auditor requirements

It is important for an auditor to have certain personal and professional attributes as well as management skills to be able to successfully plan and conduct internal audits. These attributes and skills are discussed in this chapter along with suggested requirements an internal auditor candidate should meet.

Personal Attributes of an Internal Auditor

- ◆ Open mindedness
- ◆ Sound judgement
- ◆ Analytical skills and decisiveness
- ◆ Persistence and tenacity
- ◆ Fairness and objectivity

There are many personal attributes that all auditors should possess to be successful and competent auditors. A good auditor will have a well-balanced compliment of each of these attributes. These include, but are not limited to the following:

Open Mindedness

An auditor must be able to accept interpretations of the requirements that are different than their own as long as the requirements are indeed satisfied. The auditors must also be ready to acknowledge that their perceptions are not necessarily the only correct ones. A good auditor will listen to the position presented by either another auditor or the auditee and weigh whether the position presented is valid.

Sound Judgment

A good auditor must be capable of judging whether or not an interpretation of a requirement offered by the auditee and the established system being audited satisfies the stated requirements. A good auditor is required to make decisions during the course of the audit. These decisions must be sound and made quickly to have the audit meet its objective and assist the audited department or office with a basis for improvement. The auditor must also be capable of reading situations and knowing when to press forward with an auditee and when to back off.

Analytical Skills and Decisiveness

Each auditor must be able to absorb all of the information gathered through the various forms of communication with the auditees as well as through the objective evidence gathered and deduce whether there is sufficient data to determine if the elements being audited conform to the stated requirements. Since the auditor works under stringent time constraints, quick analysis of the data and the drawing of conclusions based on this analysis is of utmost importance for a meaningful and on-schedule audit.

Persistence and Tenacity

Persistence and tenacity are probably the most important of all the internal auditor's personal attributes. An auditor needs to have the tenacity of a bulldog without having the bulldog's temperament. This means that the auditor must remain on track regardless of where the auditee tries to direct him or her without loosing his or her temper. The auditor must persist in the collection of data that is sufficient to verify whether or not the system elements conform to the established requirements. This means that the auditor must follow a line of questioning until the evidence required is obtained.

Fairness and Objectivity

An internal auditor should be willing to make allowances for the occasional human errors and mistakes when actual human errors and mistakes are uncovered. An auditor must not pick at trivial matters. The auditor, after all, is trying to assist the organization in identifying the shortcomings with respect to meeting the established requirements. For example, it would be absurd for an auditor to raise an observation or nonconformance because the organization's quality manual did not have index tabs in all the copies even though each set was complete and up to date. Unnecessary digging usually does not aid in enlisting the assistance of the auditee nor make the audit results credible in the auditee's eyes. This does not mean that the audit should be kept on the surface, but that the auditor must treat the auditee with the goal of helping the organization by being fair and objective.

Professional Attributes

> ◆ Appearance
> ◆ Punctuality
> ◆ Preparedness
> ◆ Polite and calm demeanor
> ◆ Direct communications

The credibility of an auditor often rests on the professionalism he or she exhibits during all phases of the audit. The level of professionalism demonstrated also reflects on the auditor's competence and the auditee's comfort in dealing with the auditor.

Appearance

An internal auditor projects an image with his or her appearance; therefore, the auditor must dress to present a competent and favorable impression to the auditee. Professional business attire will assist the auditor in setting a stage that signifies the formality of the audit process. It is important to remember that by being properly dressed and groomed the auditor and auditee's confidence will be increased.

Punctuality

The first impression that the auditee gets of the auditor is based on whether or not the auditor arrives as scheduled. The auditor must remember that the auditee will be setting aside their valuable time for the audit and will appreciate it if the auditor does not waste their time by being late. The auditor's punctuality also reflects on his or her enthusiasm and drive to help the company improve.

Preparedness

An auditor who is not prepared for the audit will soon realize that his or her credibility will be in doubt. The auditee and the auditor will soon become frustrated if the auditor is not properly prepared. Therefore, a good auditor will take the necessary time to plan and prepare for the audit in order to guarantee that the objectives for auditing are met.

Polite and Calm Demeanor

From the auditee's viewpoint an audit can be both an upsetting and stressful experience. The auditor's demeanor sets the stage for either a productive or confrontational and ineffective audit experience. If the auditor follows some simple rules, the auditee can become relaxed, less combative, and less defensive. These include avoiding participation in verbal altercations with the auditee, being understanding of the situation the auditee is in, smiling, having a brief conversation unrelated to the audit, but most of all being polite and calm.

Direct Communications

To achieve the objectives of the audit the auditor must pose the questions clearly, thereby avoiding long unnecessary exchanges between the auditee and auditor. The auditor must at all times remember that the audit is neither a social visit nor an inquisition. Being direct simply means asking the questions necessary to verify conformance and not going any further than necessary. Sometimes being direct means having the auditee explain things that the auditor does not understand. This should be allowed only if the explanation is needed to verify compliance. The auditor should respect that the audit is taking the auditee away from his or her normal work processes so wasting their time should be avoided.

Management Skills

♦ Time management
♦ Planning
♦ Organizing
♦ Communications

Management skills such as time management, planning, organization, and communication are important to conducting an effective and efficient audit. All auditors, whether acting as the lead auditor or not, need to master these management skills. As the scope of the audit, the size of the audit team, the size of the organization, and duration of the audit grows, these skills become more and more important.

Time Management Skills

The auditor, especially the lead auditor, must be able to manage the time spent auditing each function or element along with managing the audit process and directing the audit team. Flexibility is required when managing an audit team's time allocation, as the lead auditor never knows what will be uncovered during the audit process.

Planning Skills

A good auditor must be able to plan all of the audit activities from the preparation stage through the closure of nonconformances. This planning encompasses determining the scope, selection and assignment of auditors, verification activities, and report writing. There will be other items, which require planning that is not directly involved with conducting the audit itself. These include planning for the necessary transportation, secretarial and technical support, and accommodations in association with the audit to be conducted.

Organizational Skills

The auditor needs organizational skills to be able to collect the information and objective evidence gathered during the audit investigation and to organize it in such a manner as to provide a complete, clear, and concise audit report. Organizational skills are also required to develop a flow of the audit investigation and checklists used which are practical and useful. Additionally, a large audit requires that each auditor have the necessary checklists, schedules, note pads, nonconformance report forms, etc. A lead auditor has to be organized to make sure that everyone is properly equipped.

Exceptional Communication Skills

As discussed in Module 4, the key to auditing is communication whether it be verbal, nonverbal, or written. A good auditor must excel in these areas. Each phase of the actual audit involves communication. These communication opportunities include: the opening meeting which sets the stage for the audit, the audit investigation to verify conformance, the audit team meetings which assures that each auditor is conducting the audit in a unified manner, and the closing meeting and audit report where the results of the audit are presented.

An excellent command of communication skills is particularly necessary during the closing meeting because this is the point in the audit process where the auditor is most likely to be challenged. During the closing meeting verbal presentation, which is a summary of the audit report, nonconformances, and observations will be presented. The auditor must remember that the departments' or offices' management may be hearing the nonconformance statements for the first time and may take them as a criticism of their management system. Since no one likes to be criticized, the auditor will probably be challenged, especially by those who were not present during the audit and do not have all of the facts. Diplomacy is a talent that a good auditor must have while communicating with the auditee. The biggest benefit of excellent communication skills is that the auditee will feel that the auditor has performed a valuable service even though nonconformances were presented.

Suggested Internal Auditor Certification Requirements

◆ Classroom training and examination
◆ Practical experience
◆ Application
◆ Certification maintenance

To become a certified internal auditor it is suggested that the following requirements be met:

Course Work

The internal auditor candidate desiring certification as a certified internal aditor must attend and participate in an internal auditor certification course.

Course Examination

The internal auditor certification process should involve taking and passing a certification examination. The student has the option of not taking the examination, although without the successful completion of the examination the student should not become a certified internal auditor.

The examination should be designed to be completed in one hour and is made up of both multiple choice questions and written nonconformance statements. The passing grade for the examination is usually eighty percent (80%).

The examination papers need to be graded by the course instructors. Upon completion of the grading process each student should be informed by letter of the results.

For those students who achieve a passing grade, the course instructor should forward a certification package along with instructions for its completion to the director of quality or equivalent function. For students who do not pass the examination, an option of retaking the examination within 6 months of course completion should be given. If the re-examination is not taken within the six-month time frame, the student will be required to attend the course before the reexamination may be administered.

The results of the examination need to be kept in strict confidence and only provided to individuals with the express written consent of the student. At no time should it be made known the fact that a student did or did not pass the examination.

Upon successful completion a certificate of completion should be issued.

Practical Experience

Participation in a minimum of two internal audits as follows should be a requirement:

A)   First audit - Serve as Trainee/Second Auditor.

   Responsibilities:
   1) Participate in pre-audit planning session(s)
   2) Participate in pre-audit meeting
   3) Participate the opening meeting
   4) Participate in audit
   5) Participate the closing meeting
   6) Participate in audit report writing/distribution
   7) Participate in follow-up audit when required

B)   Second audit - Serve as Lead Auditor.

   Responsibilities:
   1) Conduct pre-audit planning session(s)
   2) Conduct pre-audit meeting
   3) Conduct the opening meeting
   4) Lead audit team
   5) Conduct the closing meeting
   6) Responsible for audit report writing/distribution
   7) Plan/conduct follow-up audit when required

While the candidate internal auditor is serving as the lead auditor, the second auditor on the audit team must be a certified internal auditor or an IQA/RAB registered lead assessor.

Certification Maintenance

Once certification has been granted, the auditor must maintain the certification by accomplishing the following:

> - The auditor should perform a minimum of two internal audits of the quality system per year including issuing an audit report.
> - The auditor should maintain an auditor log documenting the audits performed during the course of the year.

PART 3

REFERENCE MATERIAL

*Provides reference material that
is useful in establishing an effective
Internal Audit Program*

Example

Of

Internal Quality and Environmental System Audit Procedure

[Note: If ISO 14001 requirements are not desired,
they can and should not be included.]

This example was taken with permission from the Quality and Environmental System Procedures
of the American Bureau of Shipping.

QUALITY AND ENVIRONMENTAL SYSTEM PROCEDURE

Title **SAMPLE**	Revision Number: **SAMPLE** 0	Date Effective: 1 November 1998	Number: QSZ-999-99-P06
Internal Quality and Environmental System Audits	Prepared By:	Approved By:	Page: 1 of 8
Applicable To: All Operating Divisions - Worldwide			Volume:

CONTENTS

CHECK SHEETS

None

ATTACHMENTS

A. Internal Audit Notification	Rev. 0	1 Nov. 98
B. Internal Audit Procedure Flow Diagram	Rev. 0	1 Nov. 98

1.0   REFERENCES

A. Quality and Environmental System Manual Section 19
B. Storage and Retention of Controlled Documents and Quality Records, UWZ-999-99-P01
C. Corrective and Preventive Action, UWZ-999-99-P05
D. Internal Auditor Certification, UWZ-999-99-P06-W001
E. Non-Conformance Reporting, UWZ-999-99-P15
F. Confidentiality, UWZ-999-99-P20

QUALITY AND ENVIRONMENTAL SYSTEM PROCEDURE	Revision Number: *SAMPLE* 0	Date Effective: 1 November 1998	Number: QSZ-999-99-P06
Internal Quality and Environmental System Audits	Prepared By:	Approved By:	Page: 2 of 8

2.0 SCOPE

This document establishes procedures for internal audits. Audits determine the effectiveness of the quality and environmental system and verify that activity and related results comply with the procedures in force at the location. The procedures consider the requirements of the *ISO 9001 "Quality system – Model for quality assurance in design, development, production, installation, and servicing"*, and audits to company specific procedures to ensure that the operation is in compliance with the ISO 9001 requirements. The procedures also address the requirements of ISO 14001, Environmental management systems – Specification and guidance for use.

3.0 RESPONSIBILITY

Overall responsibility for internal quality and environmental system audits is held by the Corporate Continuous Improvement Steering Committees as administered by the Director of Total Quality.

The respective Quality Coordinator holds responsibility for internal quality and environmental systems audits.

Responsibility for carrying out internal quality and environmental systems audits is held by the auditor/assessor.

Responsibility for facilitating internal quality and environmental systems audits is held by the management of the audited office or department.

The management of the audited office or department holds responsibility for follow-up action.

4.0 DESCRIPTION OF THE PROCEDURE

The audit process is outlined in Attachment B, Internal Audit Procedure Flow Diagram.

4.1 Definitions

Audit - A systematic and independent examination to determine whether quality activities and related results comply with planned arrangements and whether these arrangements are implemented effectively and is suitable to achieve the stated objectives.

General Audit - An audit that addresses the general operation of a site, and addresses applicable sections of the Quality and Environmental System Manual, quality and environmental system procedures, and operating procedures and process instructions.

Surveillance Audit - An audit that addresses specific areas within the operation at a site, and addresses selected sections of the Quality and Environmental System Manual, quality and environmental system procedures, and operating procedures and process instructions.

Audit Checklist - A listing of specific items within a given area that are to be audited.

Audit Report/Checklist - A combination of audit report and associated checklist.

Finding - A statement of fact supported by objective evidence, about an process whose performance characteristics meet the definition of non-conformance or observation.

Non-conformance - Non-fulfillment of a specified requirement.

QUALITY AND ENVIRONMENTAL SYSTEM PROCEDURE	Revision Number: *SAMPLE* 0	Date Effective: 1 November 1998	Number: QSZ-999-99-P06
Internal Quality and Environmental System Audits	Prepared By:	Approved By:	Page: 3 of 8

Observation - A detected weakness which, if not corrected, may result in the degradation of product or service quality or potential negative impact of the environment.

Station - A location from which work processes are carried out and in which is maintained only in-process work records. A station's controlling office maintains records of the station's completed work processes.

Office - A location where work processes are carried out, or from which work processes are controlled, and where records of in-process and completed work are maintained.

4.2 Auditor Independence

Auditors who are independent of the office/department/function being audited shall conduct the audits.

4.3 Audit Scheduling

The Director of Total Quality with the assistance of the Quality Coordinators shall establish and maintain an audit schedule for all audits to be carried out.

The frequency of these audits shall be such to ensure the proper implementation and continued effectiveness of the quality system at all levels. Internal audits may be required, in addition to the above requirements, by the respective CISC.

When preparing the audit schedule and the assignment of internal auditors, the Director of Total Quality/Quality Coordinators shall consider the following:

- The companies' audit schedule.
- The extent of the audit (full audit or selected clause audit).
- Input from managers, as applicable.
- Results of previous audits.

The audit schedules shall be agreed to by the respective Continuous Improvement Steering Committee. The audit schedules shall be distributed to all members of the respective Continuous Improvement Steering Committee for confirmation of audit dates.

It shall be ensured that audits are conducted in accordance with the following requirements:

- Corporate Office - once a year. This internal audit shall be conducted and scheduled by the Director of Total Quality, or designee.
- All other Offices - once a year. The scheduling of these internal audits shall be coordinated with the Director of Total Quality, or designee.
- Stations - once a year. The audits of stations may be conducted as part of the audit of an office or independently.

Modification of the audit schedule shall be avoided as far as possible. If, for a valid reason, one of the parties needs to reschedule the audit, then this party shall advise the Director of Total Quality or respective Quality Coordinator and start negotiations involved to reach a consensus on another date for the audit.

If a consensus cannot be reached, the subject has to be brought to the attention of the respective CISC by the Director of Total Quality or Quality Coordinator for resolution.

QUALITY AND ENVIRONMENTAL SYSTEM PROCEDURE	Revision Number: SAMPLE 0	Date Effective: 1 November 1998	Number: QSZ-999-99-P06
Internal Quality and Environmental System Audits	Prepared By:	Approved By:	Page: 4 of 8

4.4 Selection of the Audit Team

The Director of Total Quality or Quality Coordinator shall select the audit team from an Approved List of Internal Auditors, taking into account the location of the audit, the availability of auditors, and their proximity to the audit location.

For this purpose, the Director of Total Quality or Quality Coordinator shall notify the auditors and their managers as early as possible, but normally no later than 4 weeks before the scheduled date of the audit.

Upon consulting with the manager, the auditor shall confirm availability within 1 week of receipt of the notification.

4.5 Audit Notification

Except as noted below for stations, at least 2 weeks before a scheduled audit the lead auditor shall notify the manager of the office to be audited with a copy to the respective managers (as applicable) and the Quality Coordinator (as appropriate).

The Internal Audit Notification Form (Attachment A) may be used for this purpose, or notification may be by letter, fax, or E-mail.

During scheduled office audits, associated stations may be audited without prior notification. When choosing stations to audit, the auditor shall consider the station activities during the period of the desired audit.

The manager of the office to be audited shall send a confirmation of receipt of the audit notification to the lead auditor with a copy to the respective Quality Coordinator no later than 3 business days after receiving the notification. This may be done by letter, fax, or E-mail.

4.6 Audit Preparation

A general audit of an office shall address the operation of the site in a comprehensive manner. Subsequent surveillance audits (see 4.1 Definitions) may be more limited in scope. The Director of Total Quality or Quality Coordinator shall designate audits that are general in nature.

Prior to the audit, the auditor shall review the Quality and Environmental System Manual and previous audit reports including non-conformance statements and corrective and preventive actions. Upon review of this information, the auditor shall add to the prepared checklist as necessary and prepare an audit plan for offices large enough to need one. This Audit plan shall be a separate document from the checklist. The audit plan shall be forwarded to the management of the office to be audited.

The lead auditor shall consider the type of the activities performed by the office and may conduct a pre-audit planning session with the audit team to determine the scope, objectives, methods, interview questions and checklist selection. At this meeting, previous audit reports for the office being audited should be reviewed. A pre-audit meeting is not a requirement.

4.7 Audit Opening Meeting

For the initial audit of a location, or at the discretion of an auditor, an opening meeting shall be held with key personnel involved in the audit. The lead auditor shall outline the audit scope and

QUALITY AND ENVIRONMENTAL SYSTEM PROCEDURE	Revision Number: *SAMPLE* 0	Date Effective: 1 November 1998	Number: QSZ-999-99-P06
Internal Quality and Environmental System Audits	Prepared By:	Approved By:	Page: 5 of 8

objectives and the audit performance and reporting methods. The audit plan shall be confirmed and arrangements made for a representative to be available in each audited area. If station audits are planned, the station(s) may be identified at this time, or the auditor may indicate that a station will or may be selected during the course of the office audit, and ask about availability of stations. In small offices, this meeting need not be formal.

If a formal opening meeting is held, an attendance register shall be completed. Minutes shall be recorded and become part of the audit report only if, in the opinion of the lead auditor, the minutes contain significant items.

Opening Meeting Agenda:

1. Introduce audit team members.
2. Confirm the purpose of the audit.
3. Review the audit scope, timetable, and agenda.
4. Provide time for questions.
5. Agree on a time and location for a closing meeting.
6. Introduce escorts who will accompany the auditors, and explain their responsibilities.
7. Circulate an attendance register.

4.8 Audit Execution

The lead auditor shall coordinate and direct the course and pace of the audit. The audit shall be conducted following the audit plan.

Each auditor shall use a checklist. Standard checklists that also serve as the audit report are available from the Quality Coordinator or Director of Total Quality. Objective evidence shall be examined and details necessary for clarification recorded. Every audit checklist item utilized as part of the audit shall have a comment. Specific details of non-conformances, together with any applicable references shall also be recorded. Findings and observations shall be discussed with the management representative as soon as they are established. Items, which require follow-up in other offices/departments, shall also be noted.

The auditor may elect to audit a station reporting to the office to verify proper control of the station. The station audit may be general or surveillance.

The audit may address any phase of a project, such as plan review, site verification of inspections or surveys. An auditor is encouraged to visit the site where a process is taking place to ascertain that process controls are in place and effective.

Upon completion of the audit, and before the closing meeting, the audit team (if more than one auditor is present) shall meet to evaluate the evidence generated during the audit. The team shall analyze any identified non-conformances to ensure validity as audit findings. Objective evidence of a departure from approved procedures, documented requirements or other applicable documents shall be considered as valid justification for an audit finding. All findings of non-conformance shall be recorded on a Corrective and Preventive Action Request form (UWZ-999-99-P05 Attachment A).

When the same non-conformance is found to occur in more than one department/function a separate non-conformance shall be issued to each department where the non-conformance was found. In no case shall a general non-conformance be issued covering multiple departments/functions or findings.

QUALITY AND ENVIRONMENTAL SYSTEM PROCEDURE	Revision Number: *SAMPLE* 0	Date Effective: 1 November 1998	Number: QSZ-999-99-P06
Internal Quality and Environmental System Audits	Prepared By:	Approved By:	Page: 6 of 8

A summary report with findings and observations including all non-conformances and observations shall be prepared.

4.9 Closing Meeting

The lead auditor, together with the audit team, shall hold a closing meeting with the management representative and any other personnel deemed necessary by the management representative. This closing meeting need not be a formal meeting for those offices or departments small enough not to need a formal meeting.

If a formal closing meeting is held, an attendance register shall be completed. Minutes shall be recorded and become part of the audit report only if, in the opinion of the lead auditor, the minutes contain significant items.

Closing Meeting Agenda:

1. Restate the purpose and scope of the audit.
2. Present a summary report of the audit including all findings and observations.
3. Clarify any of the findings and observations, if requested.
4. Advise attendees of the intended issue date of the formal audit report.
5. Remind local management to issue a follow-up action plan in accordance with this procedure (see 4.12 Follow-up Action).
6. Advise office management of any intention to perform a follow-up audit.
7. Circulate an attendance register.

If the lead auditor deems a follow-up audit necessary, the auditor shall advise the Director of Total Quality and respective Quality Coordinator. The lead auditor, based on the follow-up plan and actions taken may adjust this determination.

4.10 Post Audit Reporting

The lead auditor shall prepare and distribute a formal audit report as early as possible but not later than 2 weeks after completing the audit. The report is to be based on completed checklists, non-conformances and observations.

The audit shall be reported on a combination audit report/check sheet. All files reviewed and action taken during the audit shall be recorded on the check sheet or an attachment.

The list of findings shall be clearly presented at the end of the audit report to enable management to see, at a glance, the outcome of the audit.

In the event the auditor desires to change something in an audit report after the same is distributed, this may be accomplished by use of an amending letter or fax which shall be sent to the same distribution as the original. A copy of the amending document shall be retained with the audit report. For substantial revisions, the auditor may send a revised report.

The audit report shall included the following:

- restatement of the audit scope
- a statement that the auditee is required to submit an action plan within 4 weeks of receipt of the audit report
- audit checklists

QUALITY AND ENVIRONMENTAL SYSTEM PROCEDURE	Revision Number: *SAMPLE* 0	Date Effective: 1 November 1998	Number: QSZ-999-99-P06
Internal Quality and Environmental System Audits	Prepared By:	Approved By:	Page: 7 of 8

- attendance registers from opening and closing meetings (if applicable)
- auditor conclusions
- auditor recommendation
- recommendation for follow-up audit, if appropriate
- list of findings, including non-conformances and observations
- corrective action requests based on non-conformances
- any other useful attachments

4.11 Report Distribution

The audit report shall be sent to the audited office. Copies shall be sent to:

- the appropriate Quality Coordinator
- the Director of Total Quality
- appropriate Vice President responsible for the function(s) audited
- to the management of other offices responsible for the audited office, if applicable.

4.12 Follow-up Action

Within four (4) weeks of receipt of the audit report, the management of the audited office shall develop a follow-up action plan to address the audit findings (see 4.1 Definitions). The plan shall be sent to the lead auditor, the respective Quality Coordinator, the Director of Total Quality, and any other person to whom the associated audit report was sent.

Corporate auditors and Quality Coordinators conducting the audit may at their discretion waive the requirement for a follow-up action plan provided:

- no non-conformances were issued during the audit,
- only a limited number of observation were raised, and
- the actions to eliminate the observations have been implemented by the audited office within four (4) weeks of receipt of the audit report.

The Continuous Improvement Steering Committee or designee shall review the auditee's follow-up action plan and CAR recommendations and/or resolutions. For audits carried out by the Corporate office or by a Quality Coordinator, when they are the lead auditor, they shall approve the recommendations presented in the auditee's follow-up action plan. For other audits, the respective Quality Coordinator shall approve the recommendations presented in the auditee's follow-up action plan. This approval shall be in writing or by E-mail within two weeks of receipt. The Action Plan approval process does not preclude the auditee from taking corrective and preventative action as soon as a finding is known.

The head of the audited office or department shall be responsible for implementing corrective and preventive actions. A record of the closure of a Corrective and Preventive Action Request shall be provided to the respective Quality Coordinator. Once the action plan has been appropriately considered, the Quality Coordinators shall verify closure of the audits. The closure of the audits shall be reported to the respective CISC and documented in the minutes. In addition, the corrective and preventive actions shall also be verified at the next internal audit.

QUALITY AND ENVIRONMENTAL SYSTEM PROCEDURE	Revision Number: *SAMPLE* 0	Date Effective: 1 November 1998	Number: QSZ-999-99-P06
Internal Quality and Environmental System Audits	Prepared By:	Approved By:	Page: 8 of 8

4.13 Follow-up Audits

Based on the audit report and the follow-up action plan, the lead auditor, the respective Continuous Improvement Steering Committee or the Corporate Continuous Improvement Steering Committee may recommend a follow-up audit.

If it is determined through the follow-up audit that corrective and preventive action is inadequate, a new report shall be generated, attached to the original report, and forwarded to the responsible manager for response.

5.0 TRAINING AND KNOWLEDGE

Internal auditors shall be trained in accordance with Quality System Procedure, Internal Auditor Certification, UWZ-999-99-P06-W001.

6.0 QUALITY RECORDS

All audit reports are Quality Records and shall be maintained in accordance with Storage and Retention of Controlled Documents and Quality Records, UWZ-999-99-P01.

7.0 FILES

Correspondence and documentation related of an internal quality audit shall be filed by the audited office, by the respective Quality Coordinator and by the Director of Total Quality.

8.0 CONFIDENTIALITY

All documents resulting from this procedure shall be controlled by the standard confidentiality policy in Confidentiality, UWZ-999-99-P20.

9.0 REVISION HISTORY

Revision	Summary	Effective
0	Initial Issue.	1 NOV 98

Example

Of

Corrective and Preventive Action Procedure

This example was taken with permission from the Quality System Procedures of the American Bureau of Shipping

Company Name   **Quality System Procedure**

Title:	Revision Number:	Date Effective:	Number:
Corrective and **Preventive Action**	1 *(Sample)*		
	Prepared By:	Approved By:	Page: 1 of 7

Applicable To:
(State to whom it is applicable in this space)

CONTENTS

FIGURES:

Figure 1   Corrective and Preventive Action Flow Diagram   Rev. 1   1 Feb 1998
(See Figure 8-1 of Chapter 8)

ATTACHMENTS:

A.   Corrective Action Request Form (CAR)   Rev. 1   1 Feb 1998

1.0   REFERENCES

(List reference procedures here)

2.0   SCOPE

This procedure addresses the requirements for Corrective and Preventive Action.

QUALITY SYSTEM PROCEDURE **Corrective and Preventive Action**	Revision Number: 1 *(Sample)*	Date Effective:	Number:
	Prepared By:	Approved By:	Page: 2 of 7

3.0 DESCRIPTION OF THE PROCEDURE

3.1 Definitions

Corrective Action Request (CAR) - a documented request to improve a Procedure, a Work Instruction, or to initiate an internal investigation of the cause of a non-conformance for applying corrective or preventive action.

Corrective Action System (CAS) - the system for identifying, analyzing, and implementing improvements and corrective action in products, services, and work processes.

Product - the output of any process. It consists mainly of goods, software, and services.

Non-Conformance - Non-fulfillment of a specified requirement.

Opportunity for Improvement - An input to the Corrective Action System that is not initiated by a Non-conformance, but is nevertheless an opportunity for improving a Procedure, or a Work Instruction.

Corrective Action - action to eliminate the causes of an existing non-conformance, non-conforming services, or other undesirable situation.

Preventive Action - action to eliminate the causes of a potential non-conformance, non-conforming services, or other undesirable situation.

Assign - for the purposes of this Procedure, the term "assign" means to transfer responsibility, with mutual consent, for action to an individual or group (i.e., team or committee).

3.2 Initiation

See Figure 1, Corrective and Preventive Action Flow Diagram.

3.2.1 Any employee or subcontractor can initiate a CAR using the form shown in Attachment A.

3.2.2 A CAR shall be initiated to document and track corrective and preventive action as a result of:

- Client feedback
- Internal or external audit results
- Analysis of trends associated with the following:
  - Customer Data
  - Client feedback
  - Non-conformances
  - Process monitoring
- Identified improvement opportunities that involve changes to Procedures or Process Instructions that are beyond the originator's area of responsibility.
- Non-conformances.

QUALITY SYSTEM PROCEDURE	Revision Number:	Date Effective:	Number:
Corrective and Preventive Action	1 *(Sample)*		
	Prepared By:	Approved By:	Page: 3 of 7

3.2.3   The CAR initiator shall provide recommendations and shall obtain and provide as many facts as possible before initiating the CAR. When Part 1 of the CAR form is completed, with the exception of the CAR number, (Attachment A), it is sent directly to the respective Quality Coordinator. Note: If the CAR initiator is an internal auditor, the originator recommendations are to be completed by the auditee.

3.2.4   A CAR is not needed to change Procedures or Process Instructions within one's own area of responsibility. These are documented and tracked according to Quality System Procedures, Development of Procedures and Process Instructions.

3.3   Corrective and Preventive Action System

3.3.1   The intent is for corrective and preventive action to occur at the lowest possible organizational level. Overall awareness by the respective Continuous Improvement Steering Committee is provided to ensure correct and documented resolutions. Such actions are intended to address situations that require correction and identify preventive action to diminish the opportunity for non-conforming services.

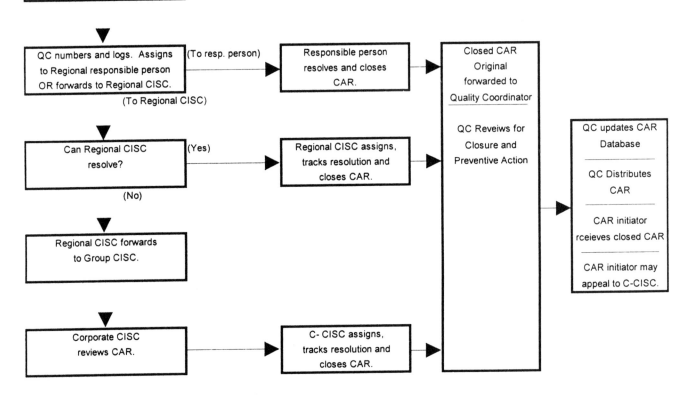

QUALITY SYSTEM PROCEDURE	Revision Number:	Date Effective:	Number:
Corrective and Preventive Action	1 *(Sample)*		
	Prepared By:	Approved By:	Page:
			4 of 7

3.3.2   The simplified flow chart (see 3.3.1 above) shows the path of a CAR through the system. The responsible Quality Coordinator shall assign a number and log the CAR into the system for tracking and trend analysis. CARs shall be numbered in accordance with the Quality System Procedure, Numbering and Indexing. A copy of the numbered CAR shall be returned to the initiator's supervisor for information.

3.3.3   A target closure date shall be shown for the Corrective Action Request. This date shall be set by the responsible authority for resolution of the CAR.

3.3.4   Corrective and preventive actions are determined from the information obtained in respect to actual and potential causes of non-conformance and the risks involved. These actions may include, among others, improvement of procedures, work instructions, training and resources.

3.3.5   Each Regional/Corporate Quality Coordinator shall maintain a status database of all CARs originated in that Regional/Corporate office. Enough detail shall be included in the database to allow for identification, traceability and trend analysis. CAR status shall be provided to the Director of Total Quality and Corporate Quality Coordinator upon request. The format shown in Attachment B shall be used.

3.3.6   The Quality Coordinator may discuss the CAR with the Regional CISC. The Regional CISC may determine the corrective and preventive action, including setting priorities and requesting resources, or it may form a team to decide the corrective and preventive action. The Regional CISC will maintain control of and be responsible for the CAR and will contact the responsible party to request assistance to resolve the CAR.

Instead of direct consideration of a CAR by the Regional CISC, the Regional CISC may instruct the Quality Coordinator to forward CARs directly to Regional personnel or to the Corporate CISC to speed resolution. If responsibility for action is in doubt, the CAR should be held for Regional CISC discussion. If the CAR is retained within the Region, the Regional CISC is responsible for tracking. If the CAR is sent to the Corporate CISC, the Corporate CISC is responsible for tracking.

If resolution cannot be reached on the Regional level, the CAR shall be forwarded to the Corporate CISC. The Corporate CISC may itself provide a solution. It may instead assess priority and assign appropriate individuals, departments, or functions to provide a solution, or it may form a team to address the CAR.

A table of functionally responsible authorities is listed here for use by the CISC.

Function	Responsible Authority
Quality System	Director TQM
Finance	F-CIT Chairperson
Human Resources	HR-CIT Chairperson
Information Management	IMS-CIT Chairperson
Operations	OPS-CIT Chairperson

QUALITY SYSTEM PROCEDURE	Revision Number:	Date Effective:	Number:
Corrective and	1 *(Sample)*		
Preventive Action	Prepared By:	Approved By:	Page:
			5 of 7

The Corporate Quality Coordinator may distribute CARs to Corporate personnel or to another CISC or to a CIT to speed resolution. If responsibility for action is in doubt, the CAR should be held for discussion by the Corporate CISC.

The Corporate CISC shall maintain control of CARs it tracks until resolved, though a CAR may have been assigned elsewhere for resolution.

3.3.7   A member of the respective CISC or CIT shall be responsible for each CAR, coordinating communication and obtaining the mutual consent for action from assignees. Should consent not be obtained, the CAR action shall revert to the responsible CISC.

3.3.8   An investigation shall be made for all CARs resulting from a non-conformance or negative client feedback. A CARs solution shall focus on eliminating the cause and preventing recurrence of the non-conformance or negative client feedback and shall be documented on the bottom of page 1 of the CAR form.

Training and resource needs shall be evaluated as part of any investigation and documented as part of "*Action(s)Taken to Resolve the CAR*" on page 2 of the CAR form.

3.3.9   Individuals and teams shall recommend resolution of root causes and corrective and preventive actions to the appropriate CISC or CIT.

3.4   Immediate Action

3.4.1   CARs considered needing immediate action shall be handled in the most direct route possible. The originator and his/her manager shall consider the risks to safety of life, property and the natural environment and decide whether a CAR needs to be handled this way. If so, the CAR should be routed directly to appropriate individual or group identified in the table in 3.3.6. The originator should propose resolution by the most direct route if the risk to the company, its clients or the public at large warrants such action. This path requires a management decision.

3.4.2   When a CAR is resolved using this path, the documentation requirements shall follow with the action taken and other pertinent information attached to the CAR.

3.5   Effectiveness Verification

3.5.1   The responsibility for follow-up to verify the effectiveness of a corrective or preventive action shall be designated as a part of the CAR resolution.

3.5.2   Follow-up to ensure corrective or preventive action implementation may involve a temporary increase in the information provided to management, measurements taken and provided to management, or arrangement for inclusion in internal audits, and shall be the responsibility of management.

3.6   Effects on other Processes and Services

The effect of the action taken to resolve a non-conformance shall be reviewed to decide whether other processes or services are affected by the same or similar non-conformance(s) and noted on page 2 of the CAR form.

QUALITY SYSTEM PROCEDURE	Revision Number:	Date Effective:	Number:
Corrective and	1 *(Sample)*		
Preventive Action	Prepared By:	Approved By:	Page:
			6 of 7

3.7   Closure

When there is evidence that corrective action has been implemented, a CAR may be closed by the:

- Corporate Continuous Improvement Steering Committee (C-CISC)
- Regional Continuous Steering Committees (R-CISC)
- Continuous Improvement Teams (HR-CIT, IMS-CIT, F-CIT)
- Director of Total Quality
- Quality Coordinator
- Corporate Responsible Person
- Supervisor of CAR Originator, or
- CAR Originator

CARs initiated to resolve audit non-conformances shall be closed out by the most appropriate function wherein the resolution resides and verified as noted in 3.5.2 above.

3.8   Final Distribution

The CAR routing shall be such that the respective Quality Coordinator shall have the opportunity to review all CARs and confirm that they have been properly addressed in the area of preventive action and closure prior to their final distribution

The respective Quality Coordinator shall do final distribution after review of the CAR for completion.

3.8.1   Opportunities for Improvement

The original of the closed out CAR with backup shall be sent to and retained by the respective Quality Coordinator. A copy of closed out CARs shall be distributed to the initiator of the CAR, to the department/office from where the CAR was initiated and to others who would need to understand or track the CAR and its resolution as determined by the person who closed out the CAR or by the respective manager. A copy of closed CARs shall be maintained by the office/Department wherein the CAR was initiated.

3.8.2   Non-conformances - Not Audit Related

The original CAR along with associated backup documentation shall be retained by the respective Quality Coordinator. Copies of closed CARs shall be distributed to the initiator of the CAR, to the department/office wherein the CAR was initiated, to individuals responsible for follow-up and verification and to others who need to understand or track the CAR and its resolution. A copy of closed CARs shall be maintained by the office/Department wherein the CAR was initiated.

3.8.3   Non-conformances Raised by Auditors

The original CAR and associated backup documentation shall be returned to and maintained by the respective Quality Coordinator. For audit-related non-conformances, copies of closed CARs shall be distributed to the Corporate Quality Coordinator, Director of Total Quality, Supervisor of the associated department/office wherein the non-conformance was raised, to individuals responsible for follow-up and verification, and to others who need to understand or track the CAR and its resolution. For internal audit-related non-conformances, copies of closed CARs shall be distributed to the lead internal auditor who raised the non-conformance, the CAR copies need not be retained by recipients.

QUALITY SYSTEM PROCEDURE	Revision Number:	Date Effective:	Number:
Corrective and Preventive Action	1 *(Sample)*		
	Prepared By:	Approved By:	Page:
			7 of 7

3.9 Appeals

If the initiator of a CAR does not agree with someone else's closure, the CAR, along with supporting documentation, may be submitted to the Corporate Continuous Improvement Steering Committee with a memorandum of explanation requesting reconsideration.

4.0 TRAINING AND KNOWLEDGE

All individuals involved with the development and resolution of corrective and preventive actions shall be familiar with the requirements detailed in this procedure.

5.0 RESPONSIBILITY

The overall responsibility for corrective and preventive action rests with Continuous Improvement Steering Committees. Quality Coordinators are responsible for coordinating the corrective action activities.

Each CAR assignee is responsible for making every reasonable effort to resolve a Corrective Action Request.

Initiators of Corrective Action Requests are responsible for describing the non-conformance, opportunity for improvement or analyzed trend to be addressed, and for providing a recommended resolution.

6.0 QUALITY RECORDS

CAR forms, CAR logs and team reports are Quality Records and shall be maintained for at least three years following Quality System Procedure, Storage and Retention of Controlled Documents and Quality Records.

7.0 FILES

Relevant background information shall be maintained with the original CAR.

8.0 CONFIDENTIALITY

This document and all documents developed in accordance with this procedure shall be controlled by the standard confidentiality policy in Quality System Procedure, Confidentiality.

9.0 REVISION HISTORY

Number	Summary	Effective
0	Initial Issue	11 July 1997

Corrective Action Request (CAR)
Improving the way we do our work

To (Quality Coordinator):	**CAR Number: ***
From (CAR Initiator):	
Date: **Branch Code:**	

Reason for Initiation:

(OI) ☐ Opportunity for Improvement of a Procedure or Process Instruction

(NC) ☐ Nonconformance   Audit No./NC No. (If applicable) _____

(CF) ☐ Client Feedback

Description:	**Source of CAR/NCR (check one)**
	Day-to day operation ☐
	Client feedback ☐
	Internal Audit ☐
	External Audit ☐
	Management review ☐
	Other ☐
	_____
	_____
	_____
Affected document and section:	

Car Initiator's (or Auditee if initiator is an Internal Auditor) **Recommendations:**

* CAR Forwarding:

This issue could not be resolved locally and is herewith forwarded to the responsible CISC for resolution. See section 3.3.6 of the procedure for additional guidance.

Sent to CISC on (date) [   ]

* QC completes shaded areas.

Investigation Results (for non-conformance or client feedback related CARs)

Corrective Action Request (CAR)
Improving the way we do our work

Action(s) taken to resolve the CAR: | Target Closure Date:

Action by: _____   **Date:** _____

Affects other processes: Yes ☐   (If yes, list them)

No ☐

Measurement of completion and requirements for follow-up: (for nonconformance and client feedback related CARs)
What measurements will be used to determine that the CAR resolution has been implemented?

_____

_____

Who is responsible for follow up action to ensure effectiveness?   _____

How is follow-up action verification of effectiveness to be accomplished?

_____

_____

Corrective Action Request Close-Out (After implementation has been completed)

Closed-out by: _____   Comments:
Date: _____

Closed-Out CAR Distribution

			Name of Individual	Date Sent
1.	CAR Initiator	Copy ___	_____	_____
2.	Supervisor/.Office file w/attach	Copy ___	_____	_____
3.	QC (w/attachments)	Original ___	_____	_____
4.	Follow-up designee (NCs only)	Copy ___	_____	_____
5.	Director, Total Quality (if the CAR was handled by CISC	Copy ___	_____	_____
6.	CIT (if the CAR was handled by the CIT)	Copy ___	_____	_____
7.	Additional information copies, as needed	Copy ___	_____	_____

Sample Internal Audit
Report/Check Sheet

Company Name
Quality System Internal Audit
Checklist and Report
Reference Procedure UWZ-999-99-P06

Regional Office
and
Reporting Functions

Column Number	Office/Stations Audited	Personnel Interviewed
1	**Regional Survey Office:** *(Location of Office)*	*(Name of Attendee)* *(Name of Attendee)*
2	**Station/Department:** *(Name of Station or Department)*	*(Name of Attendee)*
3	**Station/Department:** *(Name of Station or Department)*	*(Name of Attendee)*
4		

Audit Team/Signature(s) : *(Name of Lead Auditor)*

(Names of additional audit team members)

Audit Date(s): 03 March 1998

Name of Company
Quality System Internal Audit
Checklist and Report

Introduction

This Internal Audit Checklist/Report is based on the requirements of the *(Company Name)* Quality System. The checklist is configured in such a manner so as to allow customization by the audit team for the office or function(s) being audited.

Suggested improvements should be forwarded to the Director of Total Quality *(or equivalent)*.

Checklist/Report Configuration

This Internal Audit Checklist/Report is configured in a manner that will allow the completed document to meet the requirements of the concluding audit report and the checklist for which the audit report is based. It is designed to be completed without having to be typed or computerized. Copies can be made for distribution and stapled together, while the original remains with the audited Department or Function.

Internal Audit Function

The Function for which this specific Checklist and Report is configured is for the *(name of the office/department type for which this checklist is configured)*.

Audit Scope

The scope of the Internal Surveillance Audit is the evaluation of the operation in the Audited Office for compliance with the Documented Policies and Procedures as contained in the Quality System Manual, Quality System Procedures, and Operations Procedures and Process Instructions.

Instructions

Audit Approach

The purpose of the audit is to collect objective evidence, through interview with personnel and review of pertinent records to assess the level of compliance to the documented requirements. The approach is to be positive and construction with open dialogue.

All previous audit findings will be reviewed for closeout and noted accordingly as a new finding or as closed-out in the conclusion of this report.

Audit Items will be selected in advance by the audit team for each specific audit cycle. Not every item listed in this document will be audited each time.

Each Audited item must have a comment or note clarifying the sample size or audited details, even if there are no associated findings.

Audit Finding Follow-up Action Plan

The audited function is reminded that Procedure UWZ-999-99-P06, 3.7 calls for a Follow-up Action Plan to be submitted within 4 weeks of receipt of the audit report.

Each audited item must have a comment or note clarifying the sample size or audited details, even if there are no associated findings.

Internal Audit Categories
Check the Items and Categories to be included In this audit

Category 1	X	Quality Policy, Code of Ethics, Responsibility & Authority
Category 2	X	Organization and Responsibilities
Category 3	X	Training and Development
Category 4	X	Document Control and Reference Documents
Category 5	X	Interpretations & Instructions
Category 6	X	Correspondence Control
Category 7	X	File Management and Retention
Category 8	X	Internal Quality Audits
Category 9	X	Corrective & Preventive Action
Category 10	X	Client Feedback
Category 11	X	Supplier Assessment & Control/External Specialists
Category 12	X	Sub-Contractor Control
Category 13	X	Human Resources

Item No.	Audit Topic/Reference(s) Quality Policy, Code of Ethics, Quality System Management Quality Policy Related Responsibility Quality System **Category 1**	Column Number Enter OK/NC/OBS/X as applicable. Leave Blank if not checked.			Finding Detailed/Notes/Comments
		1 Office	2	3	
1-1	Is the Quality Policy available? QSM 2, Vol. 1	OK			The quality policy is prominently displayed.
1-2	Is the meaning of the Quality Policy understood by Management and Employees? QSM 2, Vol. 1	OK			During the internal audit, employees and management were interviewed as to their understanding of the Quality Policy and each answered in their own words, which indicated a good understanding of such.
1-3	Is the Code of Ethics available? Has it been provided to or presented to all employees? Is there visible evidence or interviewed evidence of this? QSM 2, 6.0 Vol. 1	OK			The code of ethics was verified to be available to all employees. Checked to ensure that this was the case by interviewing several employees and seeing visible evidence that it was available and had been read.
1-4	Is Management Aware of how the Quality Policy is implemented within the Organization? QSM 2, 6.0 Vol. 1	OK			Interviews with the management team specifically on this subject showed that they were aware of how the quality policy was implemented within the organization.
1-5	Are the Quality Objectives for the Organization known? QSM 3, 3.0 Vol. 1	OK			The quality objectives were known and being measured as to accomplishment within the organization.

Item No.	Audit Topic/Reference(s) Quality Policy, Code of Ethics, Quality System Management Quality Policy Related Responsibility Quality System **Category 1 - continued**	Column Number Enter OK/NC/OBS/X as applicable. Leave Blank if not checked.			Finding Detailed/Notes/Comments
		1 Office	2	3	
1-6	Does Management know their responsibilities regarding Quality? QSM 4, 4.0 Vol. 1 & QSM 7, 5.0 Vol.1	OK			Management was aware of their responsibilities regarding quality as ascertained during the interviews and audit process.
1-7	Do Management and Employees understand the Quality System Structure? QSM 5, 4.0 Vol. 1	OK			The quality system structure appears to be well understood by all personnel interviewed.
1-8					
1-9					
1-10					

Item No.	Audit Topic/Reference(s) Organization and Responsibilities Category 2	Column Number Enter OK/NC/OBS/X as applicable. Leave Blank if not checked. 1 Office	2	3	Finding Detailed/Notes/Comments
2-1	How are Organizational Responsibilities defined? (*Organization chart, Delegation of Authority*) QSM 7, 4 Vol. 1 & Figure 7-1 & 7-2 and AWZ-012-03-P03 Vol. 23	OK			Organization Chart dated 23 February 1998 defines responsibilities on the chart through added codes which define those responsibilities for each person.
2-2	Are staff Inter-relationships defined? AWZ-012-03-P03 Vol. 23	OK			Yes, through the organization chart and the position descriptions, all of which were in evidence.
2-3	Are Management personnel aware of the Structure for achieving Quality? QSM 7, 5.0 Vol. 1	OK			Yes, as ascertained through the audit interviews.
2-4	Is it known by management & employees who the Is it known who the Quality Coordinator is? QSM 7, 5.6 Vol. 1	OK			Interview with employees showed that they did know who was their respective quality coordinator and Director of Total Quality is.
2-5	Are the staff familiar with their responsibilities? Have they seen their respective Position Description? AWZ-012-03-P03 Vol. 23	OK			All individuals have seen their respective Position Descriptions

Item Now	Audit Topic/Reference(s) Organization and Responsibilities	Column Number Enter OK/NC/OBS/X as applicable. Leave Blank if not checked.			Finding Detailed/Notes/Comments
	Category 2 - continued	1 Office	2	3	
2-6	Are adequate resources provided? *(Computer, software, equipment)* QSM 2, 6.2 Vol. 1	OK			Yes, adequate computer hardware/software are in evidence.
2-7	Are human resource requirements identified? Are adequate human resources provided? AWZ-012-03-P03, 3.1 Vol. 23 & UWZ-999-99-P094, 3.1 Vol. 2	OK			Yes, adequate human resources are in evidence.
2-8					
2-9					
2-10					

Item No.	Audit Topic/Reference(s) — Training and Development — Category 3	Column Number: Enter OK/NC/OBS/X as applicable. Leave Blank if not checked. — 1 Office	2	3	Finding Detailed/Notes/Comments
3-1	Has a Training Needs Assessment been done for the office/function which includes the reporting stations? AUZ-012-03-P05-W001, Vol. 23	NC			A Training Needs Assessment has not been completed for all individuals within the office. (*List of those for whom it was not completed*).
3-2	Has an Office Training Plan been completed which includes all employees in the office/function/associated stations? AUZ-012-03-P05-W001, Vol. 23	OK			Department Training Plan reviewed dated 30 July 1997, but did not contain all personnel. See Item 3-1 above.
3-3	Is there evidence of completed training course feedback forms having been submitted to HR? UWZ-999-99-P04, 3.5 Attach A Vol. 2	OK			There was no visible evidence that the required training course feedback forms are being used and forwarded to HR, but then there has been no training either. See Item 3-5.
3-4	Have training needs been incorporated into the department budget? UWZ-999-99-P04, 3.4 Vol. 2	OBS			The budget contains moneys for training, but it was not possible to tell if there is a direct correlation between the budget and the training needs.
3-5	Is the Training Plan monitored against actual training that has taken place? UWZ-999-99-P04, 3.3 Vol. 2 & AUZ-012-03-P05-W001, Vol. 23	NC			The plan is monitored, but no training has taken place since the last internal audit.

Item No.	Audit Topic/Reference(s) Document Control Reference Documents **Category 4**	Column Number Enter OK/NC/OBS/X as applicable. Leave Blank if not checked. 1 Office	2	3	Finding Detailed/Notes/Comments
4-1	Are there any controlled document receipt acknowledgments outstanding for employees of this office, function or associated stations which are greater than six weeks old? UWZ-999-99-P02, 3.4.1, 2, Vol. 2	OK			There are none outstanding for this office.
4-2	Are any obsolete documents (not the latest revision) in use in this office/function or station? Are appropriate and current documents in use? UWZ-999-99-P02, 3.4.1 Vol. 2	OK			There were no obsolete documents discovered during examination of the documents.
4-3	Are the latest revision of forms being used in carrying out the work processes? (Check forms in use and any stock piles of forms, reports, certificates and check sheets) UWZ-999-99-P02, 3.6 Vol. 2 & SWZ-002-99-P08-W003, 2 Vol. 17	OK			The latest computer forms disk was in evidence.
4-4	Does the office/function have local procedures? If so, are they required by Law? Were they approved? (If the process is covered by an existing Work Instruction, the need should be questioned; if not needed, and OBS should be made) UWZ-999-99-P07, 3.4 Vol. 2	X			No local procedures were in evidence.
4-5	Do all employees performing work have access to copies of Procedures/Process Instructions and/or are copies issued to them? UWZ-999-99-P02, 3.4.1.1 Vol. 2 & TWZ-999-99-P16, 3.2.1 Vol. 2	OK			Copies are available at several locations within the office.

Item No.	Document Control Reference Documents — **Category 4- continued**	Column Number (Enter OK/NC/OBS/X as applicable. Leave Blank if not checked.) 1 Office	2	3	Finding Detailed/Notes/Comments
4-6	Are sub-distributed controlled documents to sub-contractors done in a controlled manner with acknowledgment receipts? UWZ-999-99-P02, 3.4.2 & 3.4.1.4 Vol. 2	OK			Sub-distributed controlled documents to sub-contractors is done in a controlled manner. Examined the files of several sub-contractors to verify this.
4-7	Are copies of Interpretations & Instructions (I/Is) sub-distributed to others deemed affected by them? Does the Station receive I/Is Direct? TWZ-003-01-P01, 3.5.1.2 Vol. 18	OK			I/Is are sub-distributed to sub-contractors in a controlled manner as evidenced by examination of the sub-contractor files.
4-8	If electronic forms are used, were they distributed properly from Corporate issued material (forms, reports, certificates)? SWZ-002-99-P08-W003, 5 Vol. 17	OK			Master disk of latest issue was in evidence.
4-9	Are working copies noted in a list placed with the originating controlled document? UWZ-999-99-P02, 3.4.1.5 Vol. 2	OK			Working copies were in evidence and those checked were also listed.
4-10	Are locally applicable documents (*Local procedures, etc.*) distributed in a controlled manner? UWZ-999-99-P02, 3.4.3 Vol. 2	X			See Item 4-4.

Item No.	Audit Topic/Reference(s) Document Control Reference Documents Category 4 - continued	Column Number Enter OK/NC/OBS/X as applicable. Leave Blank if not checked.			Finding Detailed/Notes/Comments
		1 Office	2	3	
4-11	Are controlled documents for ex-employees or sub-contractors properly collected and issuers notified? UWZ-999-99-P02, 3.4.4 Vol. 2	OK			There was no evidence to indicate that this was not being done.
4-12	Is there a distribution matrix for controlled documents? QSM 10, 4.3.2 Vol. 1	OK			A distribution list which was current was in evidence
4-13	Does a Master List exist to identify current revision levels of controlled documents? QSM 10, 4.3.3 Vol. 1	OK			The corporate distributed master list was in evidence
4-14					
4-15					

Item No.	Audit Topic/Reference(s) Interpretations/Instructions (I/Is) Category 5	Column Number Enter OK/NC/OBS/X as applicable. Leave Blank if not checked.			Finding Detailed/Notes/Comments
		1 Office	2	3	
5-1	Have any Interpretations/Instructions been received by the audited office/function/station? If yes, are they in evidence? If no, should they have them?	OK			Yes, I/Is received from Corporate and were in evidence.
5-2	Was the issued I/I's done as a controlled document? Are acknowledgments in evidence?	OK			Issued by Corporate in a controlled manner. Acknowledgments on file.
5-3	Have any sub-distribution of I/I's taken place? If so, is there evidence they were done in a controlled manner? If not, should there have been?	OK			Sub-distribution to sub-contractors handled in a controlled manner.
5-4	Is there a Master List of I/I's in evidence which has been distributed within the past 6 months?	OK			Yes, listing on controlled documents sub-distribution file.
5-5	Are any obsolete I/I's Notices in evidence?	OK			None seen during audit

Item No.	Audit Topic/Reference(s) Correspondence Control Category 6	Column Number Enter OK/NC/OBS/X as applicable. Leave Blank if not checked.			Finding Detailed/Notes/Comments
		1 Office	2	3	
6-1	Is correspondence requiring action tracked to assure response? *(Verify that the process used by the office/function meets this requirement)* UWZ-999-99-P23, 3.4 Vol. 2	OK			Correspondence tracked using 'in', 'pending' and 'out' trays.
6-2	Is E-Mail requiring action tracked to assure response? *(Look at the incoming folder)* UWZ-999-99-P23, 3.4 Vol. 2	OK			Yes, using folder system
6-3	Does project related correspondence (substantive correspondence) contain File Reference Codes, Project ID, Supplemental ID information? UWZ-999-99-P23, 3.2 Vol. 2	OK			Documentation reviewed contained identification.
6-4	Is each job (class or survey request) identified upon receipt and identification maintained, either directly or by cross-reference on all documents & records relating to the job? UWZ-999-99-P17 Vol. 2	OK			Yes, all project related material was identified upon receipt with ABSID numbers.
6-5					

Item No.	Audit Topic/Reference(s)	Column Number Enter OK/NC/OBS/X as applicable. Leave Blank if not checked.			Finding Detailed/Notes/Comments
	Category 7 File Management & Retention	1 Office	2	3	
7-1	Are there procedures for maintaining files & records in this office/function? Is P25 used or local procedures? If local, do they conflict with P25? UWZ-999-99-P25, 3.2.1 Vol. 2	OK			P25 is used.
7-2	Are the office/function maintain in-process work files organized in accordance with the requirements of P25? UWZ-999-99-P25, 3.2.2 Vol. 2	OK			In-process work files are maintained in accordance with P25, per examination of several files.
7-3	Is a list of file types retained in the office/function and is it in evidence? Has it been updated in the past year or is it still current? UWZ-999-99-P25, 3.2.3 Vol. 2	OK			Such a listing was in evidence and it was dated as current.
7-4	Is there a person designated for safe keeping of the files? UWZ-999-99-P25, 3.2.3 Vol. 2	OK			Yes, as per the organization chart delegation.
7-5	Is there an active file retrieval system in place for those offices large enough to need one (place holder cards or clipboard sign out)? UWZ-999-99-P25, 3.2.4 & 3.2.5 Vol. 2	OK			Noted file check out system used.

Item No.	Audit Topic/Reference(s) File Management & Retention Category 7 - continued	Column Number Enter OK/NC/OBS/X as applicable. Leave Blank if not checked.			Finding Detailed/Notes/Comments
		1 Office	2	3	
7-6	Does the office/function retain their own in-process work records (reports, invoices, etc.) and work completed by stations they control in an organized manner? UWZ-999-99-P25, 3.4 Vol. 2	OK			Complete Project files in evidence. Examined several.
7-7	Are the Project Files kept in accordance with the retention requirements of P25 Attachment A? UWZ-999-99-P25, Attach A Vol. 2	OK			Files are kept in accordance with P25 per examination of same.
7-8	Is File System Monitoring done? UWZ-999-99-P25, 3.3 Vol. 2	OK			File system monitoring was verified as being done during normal file access.
7-9	Does the office have a basis for retaining files beyond system requirements? *(Excessive file retention may create a space problem & consider the frequency of use versus space required)* UWZ-999-99-P25, Attach A Vol. 2	OK			Although no specific basis, the files are maintained so long as those responsible feel comfortable in their disposal.
7-10	Do Stations maintain in-process work files which are organized in accordance with the requirements of P25? UWZ-999-99-P25, 3.4 & 3.5 Vol. 2				

Item No.	Audit Topic/Reference(s) Internal Quality Audits **Category 8**	Column Number Enter OK/NC/OBS/X as applicable. Leave Blank if not checked.			Finding Detailed/Notes/Comments
		1 Office	2	3	
8-1	Has the office/function had an internal audit conducted within the past calendar year? UWZ-999-99-P06, 3.3 Vol. 2	OK			Yes, Corporate Internal Audit 30 June - 2 July 1997. IEC-30JUN97-005-71061
8-2	Did the office/function prepare an action plan within 4 weeks of the audit or were all findings closed prior to the 4-week requirement? UWZ-999-99-P06, 3.7 Vol. 2	OK			Yes, Follow-up Action Plan dated 28 July 1998
8-3	Have all findings of the audit been closed in a timely manner? Is there evidence that the action in the action plan was was carried out as proposed? Was it effective? *(Verify the past audit)* UWZ-999-99-P06, 3.7 Vol. 2	OK			Yes, only one observation which related to organization. Resolved (see item 2-1 above). This closes the previous audit as far as this function is concerned.
8-4	Does the office maintain a complete document package for the past internal audit (report, action plan, CARs, follow-up etc.)? UWZ-999-99-P06, 6 Vol. 2	OK			Yes, file seen and examined.
8-5	Did any external audits take place in this office? Was action and response done through the Corporate office? *(Review evidence)* UWZ-999-99-P06, 3.5 Vol. 2	OK			Yes, SGS Audit 26 November 1997. All findings were examined and found to be closed as far as this function is concerned.

Item No.	Audit Topic/Reference(s) Corrective and Preventive Action Non-conformance Reporting **Category 9**	Column Number Enter OK/NC/OBS/X as applicable. Leave Blank if not checked.			Finding Detailed/Notes/Comments
		1 Office	2	3	
9-1	Are CARs which were generated by the Office/function prior to 1 September 1996 closed or status known? UWZ-999-99-P05 Vol. 2	OK			No CARs raised outside of audit.
9-2	Is there any evidence that a CAR should have been initiated but was not? UWZ-999-99-P05, 3.2.2 Vol. 2	OK			There was no evidence of this. See note above. This is noted for information, as it is felt that there is opportunity for CARs outside of the audit.
9-3	Have target dates been established for all CARs? UWZ-999-99-P05, 3.3.3 Vol. 2				
9-4	Is there any evidence of non-conformances which have not been reported with actions taken using the Corrective Action System? UWZ-999-99-P15, 3.3 Vol. 2	OK			There was no evidence of this discovered during the audit.
9-5					

Item No.	Audit Topic/Reference(s) Client Feedback	Column Number Enter OK/NC/OBS/X as applicable. Leave Blank if not checked.			Finding Detailed/Notes/Comments
		1 Office	2	3	
	Category 10				
10-1	Has there been any solicited client feedback (Regional office or Corporate office)? UWZ-999-99-P22, 3.1 Vol. 2	OK			Feedback solicited September 1997.
10-2	Has there been any unsolicited client feedback expressing clients satisfaction or dissatisfaction? UWZ-999-99-P22, 3.1 Vol. 2	OK			It was stated that no negative feedback had been received since the last internal audit. There was no evidence to dispute this statement.
10-3	Has any client feedback resulted in a CAR being initiated? If not, should there have been? UWZ-999-99-P22, 3.2 Vol. 2	OK			None. See above comment.
10-4					
10-5					

Item No.	Audit Topic/Reference(s) Supplier Assessment & Control External Specialists **Category 11**	Column Number Enter OK/NC/OBS/X as applicable. Leave Blank if not checked. 1 Office	2	3	Finding Detailed/Notes/Comments
11-1	Are External Specialists' qualification data files, appraisal letters and record of annual reviews maintained in the office and are they complete? SWZ-002-99-P10, 6 Vol. 17	OK			Yes, file reviewed and up-to-date.
11-2	Have copies of all External Specialist's approval letters been sent to CDC, Houston? SWZ-002-99-P10, 3.3.6 Vol. 17	OK			Yes, verified through their being on the listing from Houston via. Safenet.
11-3					
11-4					
11-5					

Item No.	Audit Topic/Reference(s) Sub-contractor Control **Category 12**	Column Number Enter OK/NC/OBS/X as applicable. Leave Blank if not checked.			Finding Detailed/Notes/Comments
		1 Office	2	3	
12-1	Does the office maintain a current, locally controlled and approved list of sub-contracted technical personnel (non-exclusive surveyors including the processes for which they are certified? TWZ-999-99-P24, 3.4.1 Vol. 2	OK			Approved listing of sub-contractors is in existence, showing their qualifications.
12-2	Does the sub-contractor have access to or is he supplied with the necessary documents, the details of service required and the procedure/process instruction/check sheets? TWZ-999-99-P24, 3.5.1 Vol. 2	OK			Procedures/Process Instructions are issued sub-contractors was evidenced as on a controlled basis.
12-3	Are reports and certificates produced by sub-contractors examined by assigned and qualified personnel? TWZ-999-99-P24, 5.1 Vol. 2	OK			All reports and certificates are reviewed before dispatch to Client. Projects reviewed: *(List the projects reviewed).*
12-4	Is the degree of control exercised at the sub-contractors premises known? TWZ-999-99-P24, 3.5 Vol. 2	OBS			Sub-contract personnel are subject to survey activity monitoring. However, to date only one sub-contractor has been monitored. All other sub-contract personnel are accompanied by monitored sub-contractor, until such time as they themselves are monitored.
12-5	Are sub-contractors' qualification data files, appraisal letters and record or annual reviews maintained in the office and complete? TWZ-999-99-P08, 7 Vol. 2	OK			Yes. Examined several files and found all to be complete.

Item No.	Audit Topic/Reference(s) Sub-contractor Control Category 12 - continued	Column Number Enter OK/NC/OBS/X as applicable. Leave Blank if not checked. 1 Office	2	3	Finding Detailed/Notes/Comments
12-6	Is a confidentiality agreement in place? TWZ-999-99-P08, 8.0 Vol. 2	OK			A confidentiality agreement was in evidence for each of the sub-contractors examined.
12-7	Has the Vice-president approved each sub-contractor? TWZ-999-99-P08, 3.3.1 Vol. 2	OK			Yes, examined several and found each to be properly approved.
12-8	Project Files which were reviewed				(List the project files that were reviewed during the audit).
12-9					
12-10					

Opening/Closing Meeting Attendance Register

No Meeting Held ☐

Name	Opening	Closing
(List by name those who attended the opening and	X	X
closing meetings)	X	X
	X	X
	X	X
	X	
	X	X
	X	
	X	X

Auditor Conclusion:

This was a Surveillance Audit of the Regional Office – *(Location).* The audit categories listed were reviewed with two non-conformances and two observations. The non-conformances relate to training needs assessment and the taking of the planned training. The observations relate to the training budget and to the control of sub-contractors. For the most part, the Regional Office staff is "walking the talk" in so far as the quality and operating procedures are concerned. However, there are foru areas that Regional Management needs to review which involve the lack of client feedback documentation, the lack of CARs related to Opportunities for Improvement, the lack of training based on the Training Plan and the monitoring of sub-contractors.

Auditor Recommendation:

It is recommended that the *(Office Location)* Regional Management evaluate the four areas and associated findings, Items 3-5, 9-2 and 10-2 to ensure that there were indeed no opportunities for improvement, negative client feedback or necessary training that should have been taken. It is noted that the Training Procedures have been strengthened effective 1 March 98 to apply accountability to management for the accomplishment of training. In addition the Corrective Action System was modified to include strengthened preventive documentation and analysis. The Client Feedback procedure was also strengthened to provide copies of all client feedback to the Quality Coordinator for analysis and presentation to the respective CISC. These three areas will therefore be addressed in the next Internal Audit.

Distribution :

Audited Office:	X	Director of Total Quality:	X
Quality Coordinator	X	Other (specify)	X

Internal Auditor
Certification
Process
Instruction

PROCESS INSTRUCTION

Title: (SAMPLE)	Revision Number:	Date Effective:	Number:
Internal Auditor Certification	0 *(SAMPLE)*	1 November 1998	QSZ-999-99-P06-W001
	Prepared By:	Approved By:	Page: 1 of 5
Applicable To: Operating Divisions – Worldwide			Volume: 2

CONTENTS

CHECK SHEETS

None

ATTACHMENTS

A. Internal Auditor Certification Process   Rev. 1   1 Nov. 98

1.0 REFERENCES

A. Corrective and Preventive Action, UWZ-999-99-P05
B. Internal Quality and Environmental System Audits, UWZ-999-99-P06

2.0 SCOPE

To provide guidance in the certification and appointment of Internal Auditors.

These instructions apply to personnel involved in internal audits of the Quality and Environmental System.

3.0 RESPONSIBILITY

It is the responsibility of the respective Quality Coordinators to recommend candidates to be Internal Auditors.

It is the responsibility of the Director of Total Quality to ensure that all personnel certified and appointed as Internal Auditors successfully complete all phases and steps of this Process Instruction.

It is the responsibility of the Director of Total Quality to appoint, as needed, personnel certified as Internal Auditors to serve as Internal Auditor Trainers for the purpose of certifying Internal Auditors in accordance with this Process Instruction.

PROCESS INSTRUCTION	Revision Number:	Date Effective:	Number:
	0	1 November 1998	QSZ-999-99-P06-W001
Internal Auditor Certification	Prepared By:	Approved By:	Page:
			2 of 5

4.0 DESCRIPTION OF THE PROCEDURE

4.1 The Quality Coordinators shall propose persons for internal auditor training to the respective CISC. Once approved by the CISC, the auditor shall be proposed to the Director of Total Quality. The Director of Total Quality shall approve nominated candidate auditors.

4.2 Before a candidate auditor is certified as an Internal Auditor, proficiency in auditing skills (technical and communication) and knowledge of the Quality and Environmental System shall be established and verified by an Auditor Trainer.

4.2.1 Technical Skills

Technical skills include the preparation, formal reporting, and closure phases of the audit process. In addition, the candidate auditor may be required to have particular technical knowledge and skills with regard to the specific areas being audited. Technical proficiency can be assessed by the candidate's previous training, experience, and education.

4.2.2 Performance Skills

Performance skills include Audit Management skills and Communication skills required to effectively and diplomatically communicate with the auditee. Proficiency in this area shall be assessed from the candidate's performance during the On-The-Job (OJT) portion of the certification process.

4.2.3 Quality System

The candidate Auditor shall demonstrate sound knowledge of the Quality and Environmental System, including the Quality and Environmental System Manual, Quality and Environmental System Procedures, Operating Procedures, and related Work Instructions of the specific areas to be audited.

4.3 Certification Process

The candidate auditor shall complete a certification program consisting of two steps: 1) Classroom training, and 2) On-The-Job Training. The certification process is flow-charted in Attachment A.

4.3.1 Classroom Training:

Successful completion of a corporate approved Auditor/Assessor course is required. Personnel having achieved Registration as Provisional Assessors, Assessors, or Lead Assessors through the Institute of Quality Assurance, (IRQA/RAB), shall be exempt from the classroom portion. They will, however, be required to fulfill the On-The-Job portion.

4.3.2 On-The-Job Training (OJT):

Participation in a minimum of two internal audits as follows:

PROCESS INSTRUCTION	Revision Number:	Date Effective:	Number:
	0	1 November 1998	QSZ-999-99-P06-W001
Internal Auditor Certification	Prepared By:	Approved By:	Page:
			3 of 5

    A)   Serve as Trainee/Second Auditor;

Responsibilities:

1. Observe/Participate in Pre-audit planning session(s);
2. Observe/Participate the Opening Meeting;
3. Observe/Participate in Audit;
4. Observe/Participate the Closing Meeting;
5. Observe/Participate in Audit Report writing/distribution;
6. Participate in Follow-up Audit when required.

    B)   Serve as Lead Auditor[*];

Responsibilities:

1. Conduct Pre-audit planning session(s);
2. Conduct the Opening Meeting;
3. Lead Audit Team;
4. Conduct the Closing Meeting;
5. Responsible for Audit Report writing/distribution;
6. Plan/Conduct Follow-up Audit when required;

NOTE: Commencement of the Lead Auditor portion of the training process shall be based on the recommendation of the trainer for step (A) of the On-the-Job Training. While the candidate auditor is serving as the Lead Auditor, the Second Auditor in the audit team shall be an Auditor Trainer.

4.3.4 The actual number of audits to be conducted for each candidate auditor may be varied based on the assessment of the Auditor Trainer.

4.3.5 The Candidate Auditor's ability to conduct and lead an audit shall provide the basis for the Auditor Trainer's decision as to whether the candidate is ready to be certified as an Internal Auditor. If ready, the Auditor Trainer shall nominate the candidate for appointment as an Internal Auditor. If the candidate is not ready, he or she shall be given a written plan to enhance his or her abilities in specific areas. This may be specific type training such as communication skills or continued experience as a trainee/second auditor. These candidates may re-enter the certification process when they have enhanced their abilities in those noted areas. If the Auditor Trainer determines that the candidate does not possess or may not achieve the communication skills required, the certification process is then deferred to the Director of Total Quality.

PROCESS INSTRUCTION	Revision Number:	Date Effective:	Number:
	0	1 November 1998	QSZ-999-99-P06-W001
Internal Auditor Certification	Prepared By:	Approved By:	Page:
			4 of 5

5.0 TRAINING & KNOWLEDGE

All personnel appointed as Audit Trainers shall have met the requirements of this Process Instruction.

6.0 FILES

All correspondence and documentation related to the Certification of Internal Auditors shall be filed at the office of the respective Quality Coordinator. The Director of Total Quality shall maintain a consolidated file.

7.0 CONFIDENTIALITY

All documents resulting from this procedure shall be controlled by the standard confidentiality policy in Confidentiality, UWZ-999-99-P20.

8.0 REVISION HISTORY

Revision	Summary	Effective
0	Initial Issue.	1 NOV 98

**PROCESS INSTRUCTION
TRAINING CHECKLIST**

_____   _____

Employee Name   **Employee Number**

Classroom Training

The employee has successfully completed a corporate approved Auditor/Assessor course.

On - The - Job Training

The employee has served as Trainee/Second Auditor in internal audit(s), as demonstrated to my satisfaction.

The employee has served as Lead Auditor in internal audit(s) as demonstrated to my satisfaction.

The employee has demonstrated technical proficiency in the preparation, formal reporting, and closure of the audit process to my satisfaction.

Quality and Environmental Standards

The employee has sound knowledge and understanding of the Quality System, including the Quality and Environmental System Manual and the Quality and Environmental System Procedures as demonstrated to my satisfaction.

Comments:

_____   _____

Trainer's Signature   **Date**

Date.) (Enter into the TDS as the employee's certification date)_____

Trainer's Employee Number _____

Trainer's Date of Certification in this process _____

INSTRUCTIONS: By signing and dating this form the auditor trainer has verified the completion of each requirement. When all requirements have been fulfilled and the form dated and signed, the auditor trainer is to make a copy and forward it, along with a copy of the report for the specific audit used in the training, to the Director of Total Quality.

Sample
Opening Meeting
Script

Sample Opening Meeting Script

Good morning ladies and Gentlemen. I would like to introduce myself; my name is (name). I will be the lead auditor for this internal audit. The other members of the audit team are Ms. (name), Mr. (name), and Mr. (name).

Let me tell you a little about myself. I have been with XYZ Corporation for 15 years in various capacities. When I started with this company I worked in the widget design office. This assignment lasted 4 years and then I was moved as a foreman to the manufacturing operation for 9 years. For the past two years I have been involved in development and implementation of the quality system. During this time I have conducted 47 internal audits and numerous supplier audits.

Now I would like to invite the other auditors to briefly describe their backgrounds.

While we are listening to the other auditors speak I will be passing around an attendance register. Please fill it out and pass it along so that everyone has a chance to fill-in their name, title, and department.

Thank you for sharing your backgrounds with us. Now we will move along to discuss the methodology employed in this audit. In previous visits to this site the auditors have conducted prescriptive audits which have helped implement the quality system. Since, the system has been up and running for two years now, we will shift the focus of the audit to a compliance audit. This means we are here to verify compliance and identify opportunities where you can improve the system and the way you work. Unlike the prescriptive audit the auditors in a compliance audit do not tell you how to fix problems, they just identify the problem areas. In particular this audit will include a comprehensive investigation of all of the system elements related to the manufacturing operation at this site and with the interfaces between manufacturing and other departments.

The audit is scheduled to last three days. I have passed out to each of you a copy of the schedule. Lets go over it to see if you need us to make any changes or if we should proceed as it stands. As you can see each afternoon we have set aside time for the audit team to meet and discuss the activities of the day. Additionally, every morning at 7:30 am we have a debriefing meeting scheduled. Each of you should attend this meeting where we will present to you the findings of the previous day. We generally have these debriefing meetings to prevent us from having a very long closing meeting. We have used this method before and find it very successful.

During the conformance verification it is inevitable that we will find opportunities for improvement. These opportunities will be categorized either as nonconformances or observations. I would like to take the time to explain each category to you now so that it will be clear to everyone what we are presenting. (*take this opportunity* to *define what nonconformances and observations are*).

There are some requirements that we have for you. We will need a meeting room each day for the audit team meetings and debriefing meetings, and have access to a fax machine, and telephone. If you can please make these arrangements for us it would be appreciated.

I would like to take this opportunity to answer any questions you may have. (*give as much time as needed, but don't let this time extend into auditing time*)

I think we are ready to begin. Thank you for your time.

REFERENCES

Murphy, Raymond J. *Implementing and ISO 9000-Based Quality System:* Government Institutes, Inc., Rockville, Maryland, 1998.

Glossary

To effectively communicate with each other we need to have a common vocabulary and understanding of the meaning of the words we use. Like all other fields, the auditing process has generated its own vocabulary and different meaning for words used in our field. However, words are only as meaningful as the degree of comprehension attained. This section defines may of the terms used in the previous chapters. It is intended as a step towards getting everyone to speak the same language.

ASSESSMENT - A term that is synonymous with audit, sometimes used to indicate a less formal means of measuring and reporting than the normal audit. An assessment is usually limited in scope.

AUDIT - A systematic and independent examination to verify whether the auditees' quality system and related results comply with planned arrangements and requirements, and whether these arrangements and requirements have been implemented effectively and are suitable to achieve the quality system objectives.

AUDITEE - The organization, department, or function being audited. This term may also be used to refer to the department manager/supervisor being audited.

AUDITOR - A trained individual who verifies whether an organization complies with its established requirements and determines whether the organization's requirements are implemented and effective in meeting the objectives of the requirements.

AUDIT PROGRAM - The organizational structure, commitment, documented methods used to plan and perform audits.

AUDIT TEAM - The group of individuals (usually consisting of trained auditors or auditor trainees) conducting an audit under the direction of a team leader or lead auditor.

CALIBRATION - The comparison of measuring and test equipment of unknown accuracy to a measurement standard of known accuracy in order to detect, correlate, report, or eliminate by adjustment any variation in the accuracy of the instrument being compared.

CERTIFICATION - The process by which an authorized body determines, verifies, and attests to the conformance of personnel, processes, procedures, or products and services to the applicable requirements in writing.

CHARACTERISTIC - Property that helps to identify or to differentiate between entities and that can be described or measured to determine conformance or nonconformance to requirements.

CHECKLIST - A listing of statements or questions that identifies each element or area the audit is intended to address.

CLIENT - The person or organization requesting the audit. Depending on the circumstances, the client may be the auditing organization, the auditee, or a third party. For internal audits the client is usually the Quality Steering Committee or similar body responsible for the organization's quality system.

CLOSING MEETING - The meeting at the end of the audit between the auditors and the auditees representative, at which time audit nonconformances, findings, and observations are presented.

CONFIRMATION - The agreement of data or information obtained from two or more different sources.

CONFORMANCE or COMPLIANCE - An affirmative indication or judgment that a product, service, or system element has met the requirements of the relevant specifications, contract, or regulation. Also the state of meeting the requirements.

CONTINUOUS IMPROVEMENT STEERING COMMITTEE – This committee consists of management personnel with executive responsibility as well as management personnel with functional responsibility. The purpose of the CISC is to manage the overall quality system. This committee is often called Quality Steering Committee (QSC).

CONTRACTOR - Any organization under contract to furnish items or services, such as a vendor, supplier, subcontractor, fabricator, and sub-tier levels of these, where appropriate.

CORRECTIVE ACTION - Measures taken to eliminate the root cause(s) of a nonconformity, a deviation from the established requirements, or an existing condition that may adversely effect the quality of the products or services provided by the organization and to prevent its recurrence.

DEVIATION - A nonconformance to or departure of a characteristic from specified product, process, or systems requirements.

ELEMENT - Any individual component of an organization's quality system (i.e., document control or corrective action).

EVALUATION - The act of examining a process or group to established requirements and forming conclusions as a result.

EXAMINATION - A measurement to determine conformance to some specified requirement.

FINDING - The documented objective evidence that identifies a noncompliance in sufficient detail to enable corrective action to be taken by the auditee.

FOLLOW-UP AUDIT - A targeted audit that verifies that corrective action has been accomplished as scheduled, and determines that the action was effective in preventing a recurrence.

GUIDELINES - Documented instructions that are considered good practice but that are not mandatory.

INSPECTION - Activities (such as measuring, examining, or testing) that gauge one or more characteristics of a product or service and the comparison of these with specified requirements to determine conformity.

LEAD AUDITOR - The individual qualified to organize and direct an audit, report audit findings, and evaluate proposed corrective actions. A lead auditor may also be called an audit team leader.

NONCONFORMITY or NONCONFORMANCE - A departure of a quality characteristic from its intended level or state that occurs with a severity sufficient to cause an associated product or service not to meet a specified requirement.

OBJECTIVE EVIDENCE - A documented statement of fact or other record pertaining to the quality of an item or activity based on information, observations, measurements, records, or tests that can be verified.

OBSERVATION - A weakness detected in an element in the quality system that, if not corrected, may result in a degradation of the product or service quality and possibly become a nonconformance.

OPENING MEETING - The introductory meeting between the auditors and the auditee's representative, at which time the overview of the planned audit and methodologies employed is presented.

PROCEDURE - A document that specifies the requirements to perform an activity.

PROCESS - The particular method of producing a product or service, generally involving a number of steps or operations.

PRODUCT - A piece of goods manufactured for a customer or service delivered to a customer.

QUALITY MANUAL - A document stating the quality policy, quality system, and quality practices of an organization.

QUALITY POLICY - The overall intentions and direction of an organization regarding quality, as formally expressed by top management.

QUALITY SYSTEM - The organizational structure, responsibilities, procedures, processes, and resources for implementing quality management.

ROOT CAUSE - A fundamental deficiency that results in a nonconformance and must be corrected to prevent recurrence of the same or similar nonconformance.

SPECIFICATION - The document that prescribes the requirements with which the product or service must conform.

STANDARD - The documented result of a particular standardization effort approved by a recognized authority.

TESTING - A means of determining the capability of an item to meet specified requirements by subjecting the item to a set of physical, chemical, environment, or operating actions and conditions.

TRACEABILITY - The ability to trace the history, application, or location of an item(s) or activity(ies) by means of recorded identification.

VERIFICATION - The act of reviewing, inspecting, testing, checking, auditing, or otherwise establishing and documenting whether items, processes, services, or documents conform to specified requirements.

Government Institutes Mini-Catalog

PC #	ENVIRONMENTAL TITLES	Pub Date	Price
629	ABCs of Environmental Regulation: Understanding the Fed Regs	1998	$49
627	ABCs of Environmental Science	1998	$39
585	Book of Lists for Regulated Hazardous Substances, 8th Edition	1997	$79
579	Brownfields Redevelopment	1998	$79
4088 ⊙	CFR Chemical Lists on CD ROM, 1997 Edition	1997	$125
4089 ▭	Chemical Data for Workplace Sampling & Analysis, Single User Disk	1997	$125
512	Clean Water Handbook, 2nd Edition	1996	$89
581	EH&S Auditing Made Easy	1997	$79
587	E H & S CFR Training Requirements, 3rd Edition	1997	$89
4082 ⊙	EMMI-Envl Monitoring Methods Index for Windows-Network	1997	$537
4082 ⊙	EMMI-Envl Monitoring Methods Index for Windows-Single User	1997	$179
525	Environmental Audits, 7th Edition	1996	$79
548	Environmental Engineering and Science: An Introduction	1997	$79
643	Environmental Guide to the Internet, 4rd Edition	1998	$59
560	Environmental Law Handbook, 14th Edition	1997	$79
353	Environmental Regulatory Glossary, 6th Edition	1993	$79
625	Environmental Statutes, 1998 Edition	1998	$69
4098 ⊙	Environmental Statutes Book/CD-ROM, 1998 Edition	1997	$208
4994 ▭	Environmental Statutes on Disk for Windows-Network	1997	$405
4994 ▭	Environmental Statutes on Disk for Windows-Single User	1997	$139
570	Environmentalism at the Crossroads	1995	$39
536	ESAs Made Easy	1996	$59
515	Industrial Environmental Management: A Practical Approach	1996	$79
510	ISO 14000: Understanding Environmental Standards	1996	$69
551	ISO 14001: An Executive Report	1996	$55
588	International Environmental Auditing	1998	$149
518	Lead Regulation Handbook	1996	$79
478	Principles of EH&S Management	1995	$69
554	Property Rights: Understanding Government Takings	1997	$79
582	Recycling & Waste Mgmt Guide to the Internet	1997	$49
603	Superfund Manual, 6th Edition	1997	$115
566	TSCA Handbook, 3rd Edition	1997	$95
534	Wetland Mitigation: Mitigation Banking and Other Strategies	1997	$75

PC #	SAFETY and HEALTH TITLES	Pub Date	Price
547	Construction Safety Handbook	1996	$79
553	Cumulative Trauma Disorders	1997	$59
559	Forklift Safety	1997	$65
539	Fundamentals of Occupational Safety & Health	1996	$49
612	HAZWOPER Incident Command	1998	$59
535	Making Sense of OSHA Compliance	1997	$59
589	Managing Fatigue in Transportation, *ATA Conference*	1997	$75
558	PPE Made Easy	1998	$79
598	Project Mgmt for E H & S Professionals	1997	$59
552	Safety & Health in Agriculture, Forestry and Fisheries	1997	$125
613	Safety & Health on the Internet, 2nd Edition	1998	$49
597	Safety Is A People Business	1997	$49
463	Safety Made Easy	1995	$49
590	Your Company Safety and Health Manual	1997	$79

Government Institutes

4 Research Place, Suite 200 • Rockville, MD 20850-3226
Tel. (301) 921-2323 • FAX (301) 921-0264
Email: giinfo@govinst.com • Internet: http://www.govinst.com

Please call our customer service department at (301) 921-2323 for a free publications catalog.

CFRs now available online.
Call (301) 921-2355 for info.

GOVERNMENT INSTITUTES ORDER FORM

4 Research Place, Suite 200 • Rockville, MD 20850-3226
Tel (301) 921-2323 • Fax (301) 921-0264
Internet: http://www.govinst.com • E-mail: giinfo@govinst.com

3 EASY WAYS TO ORDER

1. Phone: **(301) 921-2323**
   Have your credit card ready when you call.

2. Fax: **(301) 921-0264**
   Fax this completed order form with your company purchase order or credit card information.

3. Mail: **Government Institutes**
   4 Research Place, Suite 200
   Rockville, MD 20850-3226 USA
   Mail this completed order form with a check, company purchase order, or credit card information.

PAYMENT OPTIONS

❑ **Check** (payable to Government Institutes in US dollars)

❑ **Purchase Order** (This order form must be attached to your company P.O. Note: All International orders must be prepaid.)

❑ **Credit Card** ❑ VISA ❑ MasterCard ❑ AMERICAN EXPRESS

Exp. ___ / ___

Credit Card No. _____

Signature _____

(Government Institutes' Federal I.D.# is 13-2695912)

CUSTOMER INFORMATION

Ship To: (Please attach your purchase order)

Name: _____

GI Account # (7 digits on mailing label): _____

Company/Institution: _____

Address: _____
(Please supply street address for UPS shipping)

_____

City: _____ State/Province: _____

Zip/Postal Code: _____ Country: _____

Tel: ( ) _____

Fax: ( ) _____

Email Address: _____

Bill To: (if different from ship-to address)

Name: _____

Title/Position: _____

Company/Institution: _____

Address: _____
(Please supply street address for UPS shipping)

_____

City: _____ State/Province: _____

Zip/Postal Code: _____ Country: _____

Tel: ( ) _____

Fax: ( ) _____

Email Address: _____

Qty.	Product Code	Title	Price

❑ **New Edition No Obligation Standing Order Program**

Please enroll me in this program for the products I have ordered. Government Institutes will notify me of new editions by sending me an invoice. I understand that there is no obligation to purchase the product. This invoice is simply my reminder that a new edition has been released.

Subtotal _____

MD Residents add 5% Sales Tax _____

Shipping and Handling (see box below) _____

Total Payment Enclosed _____

Within U.S:	Outside U.S:
1-4 products: $6/product	Add $15 for each item (Airmail)
5 or more: $3/product	Add $10 for each item (Surface)

15 DAY MONEY-BACK GUARANTEE

If you're not completely satisfied with any product, return it undamaged within 15 days for a full and immediate refund on the price of the product.

SOURCE CODE: BP01